The Legacy of Carbon Dioxide

Past and Present Impacts

The Legacy of Carbon Dioxide

Past and Present Impacts

Paul J. Karol

Professor Emeritus of Chemistry

Carnegie Mellon University

CRC Press

Taylor & Francis Group

Boca Raton London New York

CRC Press is an imprint of the
Taylor & Francis Group, an **informa** business

CRC Press
Taylor & Francis Group
6000 Broken Sound Parkway NW, Suite 300
Boca Raton, FL 33487-2742

© 2019 by Taylor & Francis Group, LLC
CRC Press is an imprint of Taylor & Francis Group, an Informa business

No claim to original U.S. Government works

Printed on acid-free paper

International Standard Book Number-13: 978-0-367-19134-4 (Hardback)
International Standard Book Number-13: 978-0-367-19080-4 (Paperback)

Library of Congress Cataloging-in-Publication Data

Names: Karol, Paul J., author.
Title: The legacy of carbon dioxide : past and present impacts / Paul J.
Karol.
Description: Boca Raton : CRC Press, Taylor & Francis Group, 2019. | Includes
bibliographical references.
Identifiers: LCCN 2019004600| ISBN 9780367191344 (hardback : alk. paper) |
ISBN 9780367190804 (pbk. : alk. paper)
Subjects: LCSH: Atmospheric carbon dioxide. | Atmospheric chemistry. |
Geochemistry. | Carbon dioxide. | Carbon--Isotopes.
Classification: LCC QC879.8 .K37 2019 | DDC 546/.6812--dc23
LC record available at https://lccn.loc.gov/2019004600

Visit the Taylor & Francis Web site at
http://www.taylorandfrancis.com

and the CRC Press Web site at
http://www.crcpress.com

Contents

Preface

The sphinx, Apollo 13, the White Cliffs of Dover, mass extinctions, stomach acid, and killer lakes in Africa: what do these all have in common? It is the legacy of carbon dioxide.

Motivation to assemble the assorted roles of carbon dioxide in our lives and surroundings was a personal quest for the author. It was catalyzed by a desire to come to terms with one of today's hot issues, pardon the pun: global climate change. Over the past years, I have read what proved to be countless scientific articles and media opinion pieces on global climate change and human influence on climate. Arguments from opposing sides could at times then seem equally convincing, which proved frustrating. I decided to educate myself further and found the topic and the paths down which my pursuit led me to be more and more fascinating. Yet I was not overcome with any obvious decision as to where the future of climate change would likely bring us. Mostly to put my thoughts in order and to try to complete my understanding of many related natural and human-generated phenomena, I have authored this book and titled it in as an unencumbered way as possible, *The Legacy of Carbon Dioxide*.

The Oxford English Dictionary regards *legacy* as anything handed down by an ancestor or predecessor. For the sake of simplicity, I have taken certain liberties with terminology and with quantitative expressions. For example, I ignore the slight difference between weights expressed as tons or tonnes. I use % changes rather than the geophysicists' preferred ‰. I apologize to the many experts in the field for being incomplete in my coverage and also, probably, error prone since the literature, overwhelming as it is, changes on almost a daily basis. My objective is to allow each reader to recognize how startlingly complex the entire issue of carbon dioxide is, how many different and unexpected ways carbon dioxide affects us, how we measure these effects, and how various are the degrees of certainty or uncertainty in what is currently understood.

Trying to keep up to date with studies involving carbon dioxide has proven to be a challenge. Here, the annual publication rate of scientific articles with either "carbon dioxide" or "CO_2" in their title is shown as a function of time. The points along the bottom flat curve correspond to scientific articles with "carbon monoxide" in their title for comparison, averaging a nearly constant four publications per year over the last half century. Data was extracted using the Google Scholar search engine.

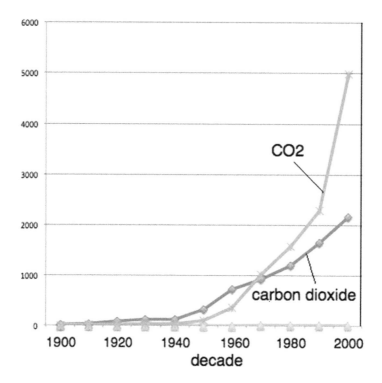

I have assiduously eschewed what prompted my interest, modern human influence on global climate, hoping each reader will reach his or her own conclusion from an unbiased presentation; a worthy goal I think – one that spites the observation of a famous author.

> There is something fascinating about science. One gets such wholesale returns of conjecture out of such a trifling investment of fact.

MARK TWAIN

> Study the past if you would define the future.

CONFUCIUS

> Every effort has been made to keep this book lean so as to encourage its reading.

Acknowledgment

One source of inspiration, advice, and encouragement and one source of invaluable assistance constitute this brief recognition of support. Meryl H. Karol, Professor Emerita of Environmental Toxicology at the University of Pittsburgh's Graduate School of Public Health and my wife, best friend, and companion for over 50 years, read most of the manuscript in its various incarnations and made countless constructive suggestions. The Mellon Institute Library staff at Carnegie Mellon University provided priceless, expedient, and instrumental assistance in accessing scientific literature crucial to moving this project along and keeping it current. Without either of these resources, my mission was doomed. Thank you.

PAUL J. KAROL

Author

Paul J. Karol is a linear academic descendent of Joseph Black, the discoverer of carbon dioxide. Karol has been on the chemistry faculty at Carnegie Mellon University in Pittsburgh for over 40 years and has received two awards for teaching during that period. His undergraduate degree in chemistry was from Johns Hopkins University and his postdoctoral research was done at Brookhaven National Laboratory. His doctorate degree in nuclear chemistry was acquired at Columbia University under the auspices of Dr. J. M. Miller. Prof. Karol has served as Chair of the Division of Nuclear Chemistry and Technology of the American Chemical Society, as Chair of the Committee on Nomenclature, Terminology and Symbols of the American Chemical Society, as Chair of the Committee on Nuclear and Radiochemical Analysis of the International Union of Pure and Applied Chemistry, and as Chair of the Joint Working Party on the Discovery of New Elements of the International Union of Pure and Applied Chemistry and the International Union of Pure and Applied Physics. He served as Visiting Professor to the Institute of Nuclear Physics in Legnaro, Italy, and the Japan Atomic Energy Research Institute in Tokai, Japan.

Academic Genealogy

During my efforts in compiling this book, I learned that I am a direct academic descendant of Joseph Black, the person credited with the characterization, if not the discovery, of carbon dioxide in 1754. This academic genealogy is shown below.

Julian M. Miller, Professor of Chemistry at Columbia University in New York City, did research in the field of nuclear chemistry. His doctoral advisor was Richard Dodson.

Richard Dodson was the first chairman of the Chemistry Department at Brookhaven National Laboratory and received his PhD from Johns Hopkins University under the direction of Prof. Robert D. Fowler.

Robert D. Fowler was on the faculty of Johns Hopkins University in Baltimore and was awarded his doctoral degree in chemistry from Willard at the University of Michigan.

Hobart Hurd Willard from Erie, Pennsylvania, received his PhD in 1909 at Harvard. The title of his thesis was "A Revision of the Atomic Weights of Silver, Lithium, and Chlorine." His studies focused on analytic methods and inorganic chemistry, particularly that of perchloric and periodic acids and their salts. Willard was a member of the chemistry faculty at the University of Michigan. His doctoral thesis advisor was T. W. Richards.

Theodore William Richards was the first American to receive the Nobel Prize in Chemistry. Richards studied at Harvard, taking as his dissertation topic the determination of the atomic weight of oxygen relative to hydrogen. His doctoral advisor was Josiah Parsons Cooke. Richards returned to Harvard as an assistant in chemistry, then instructor, assistant professor, and finally full professor.

Josiah Parsons Cooke from Boston, Massachusetts, a member of the National Academy of Sciences and president of the American Academy of Arts and Sciences published over 40 papers in his career. He received his AB in chemistry from Harvard in 1848. I. Bernard Cohen, the distinguished historian, describes Cooke as "the first university chemist to do truly distinguished work in the field of chemistry" in the United States. Cooke's education at Harvard was under the mentorship of Benjamin Silliman, although it is reported that Cooke was largely self-taught.

Benjamin Silliman, a chemist with a degree in law as well with a deep interest in geology and mineralogy, Silliman travelled extensively and became a professor at Yale despite no formal training in natural science. His influence by Thomas Hope led him to research in chemistry.

Thomas Charles Hope was Professor of chemistry and medicine at Scotland's University of Edinburgh. He was the discoverer of the element strontium in 1791. Among his students, besides Benjamin Silliman, was Charles Darwin. Hope was mentored by Joseph Black in Edinburgh.

Joseph Black, Scottish physician and chemist, was Professor of chemistry at the University of Glasgow and then the University of Edinburgh. Besides his pioneering work on carbon dioxide, he also did some foundational work on thermodynamics and invented a precision analytical balance.

1 Starting Elements

> The point of philosophy is to start with something so simple as not to seem worth stating, and to end with something so paradoxical that no one will believe it.
>
> **BERTRAND RUSSELL**
> (*The Philosophy of Logical Atomism*)

Without carbon dioxide, nature would be very different, unrecognizable. Chief among its roles is as feedstock for photosynthesis in the plant kingdom. Where did carbon dioxide originate? What better place to start? Let us go for a brief visit back to the "start" according to current scientific understanding.

NUCLEOSYNTHESIS

As the current concept of the universe's evolvement goes, about one second after the "big bang," the expanding universe was peppered with elementary particles including protons, neutrons, and electrons. Seconds later, deuterium (heavy hydrogen) and helium formed by cascades of fusion encounters, first between protons, and then with fusion residues. After a few minutes, three-quarters of the universe's matter* was hydrogen and the rest essentially helium. Traces of deuterium and other light species were present as well. From this simple mixture, the earliest stars slowly condensed through gravitation. The consequent gravitational heating within stars opened pathways to the production of heavier and heavier combinations of protons and neutrons and the elements associated with them. Despite the current age of the universe at more than some dozen billion years, its elemental composition is still 99% hydrogen and helium, implying that nucleosynthesis has not proceeded very far yet…on the average. Local variations are quite interesting though.

From the Nobel Prize winning efforts of Alsatian-born, German-educated physicist Hans Bethe, we know much about the reactions that fuel the stars and about the concomitant production of heavier and heavier elements.

---

* Exclusive of *dark* matter.

The earliest stars could sustain the fusion reactions associated with their abundant hot hydrogen by a sequence of steps in which two protons could first meld together to form a deuterium nucleus (proton plus neutron) accompanied by the release of a positively charged electron (called a positron) and some energy. A nucleus consists of positively charged protons – the number of which determines the element's identity – plus a number of neutral neutrons. The deuterium could then fuse with another proton to form a new nucleus with two protons and a neutron. The element with two protons in the nucleus is helium and this particular product with a total of *three* nuclear components is called helium-3. Its formation is accompanied by the release of still more energy. As the amount of helium-3 accumulates in the blazing hot mixture, the small probability of two helium-3s fusing becomes rather influential. Two helium-3s produce one helium-4, freeing two protons and a large burst of energy. Helium-4 has two protons in it, which is why it is helium. Its distinction from helium-3 is that helium-4 has two neutrons for a total of *four* component particles in contrast to the three for helium-3. Nuclei that have the same number of protons but different numbers of neutrons are called *isotopes* of the element with the given proton number.

The net result of these nuclear reactions is that four protons end up producing one helium-4 plus some lighter particles and energy. Helium-4 is a particularly stable combination of particles. That stability accounts both for the enormous amount of energy released in ^{4}He formation and for the reciprocal difficulty of breaking it apart. The stability is also the reason that it is the dominant reaction product present in the proton-rich material, the deuterium and helium-3 being present as minute components.

As the star matures and hydrogen is consumed by fusion to form denser, heavier helium, gravitational collapse can resume, heating up the mixture with its significant component of helium-4. Under the right conditions, a helium-burning cycle may commence. Here, two ^{4}He nuclei can fuse to form a ^{8}Be nucleus. Beryllium-8 has four protons (making it, thereby, beryllium) and four neutrons. But ^{8}Be happens to be a very unstable system. On its own, ^{8}Be breaks back down to two helium-4s in a very, very short time, less than a millionth of a billionth of a second. However, when there is abundant helium-4 nearby, it becomes possible to fuse the beryllium-8 with another helium-4 before the beryllium has split up. This second step in the helium-burning cycle produces carbon-12.

$$^8\text{Be} + {}^4\text{He} \rightarrow {}^{12}\text{C}$$

As the mixture continues to brew, carbon-12 will have the right conditions to coalesce with yet another of the surrounding helium-4s producing oxygen-16, a nucleus with eight protons and eight neutrons. That, in brief, is the origin of the carbon and oxygen atoms that will link up to make our carbon dioxide molecular combination. Carbon and oxygen are the third and fourth most abundant elements in the universe.

THE ELEMENTS OF CHEMISTRY

Everyone has probably heard the expression "opposites attract." This is rigorously the case with electrostatic charges. The positive nucleus has a strong attraction for negative particles which are present as electrons. If the number of electrons around the nucleus equals the number of protons in the nucleus, then the combination is electrically neutral since the amount of charge on a proton is equal to that on an electron, but just opposite in sign. The nucleus with its entourage of electrons forms a neutral species called an *atom*. An atom, by definition, is uncharged. An atom of carbon has a total of six electrons and an atom of oxygen has a total of eight electrons.

The total mass of an atom is that of the nucleus plus that of the very light electrons, less an extremely small amount that is released when the electrons become bound about the nucleus. The atomic mass can be determined very precisely, relative to some standard, using modern instruments. The mass standard adopted by all scientists is carbon-12, whose atomic mass is arbitrarily, but very conveniently, set at exactly 12 "mass units," *u*. Atomic masses of some important isotopes of both carbon and oxygen are listed in the table along with the percentage of the isotope found in naturally occurring terrestrial samples.

Isotope	Mass*	Abundance
Carbon-12	12.0000000 u	98.90%
Carbon-13	13.0033548 u	1.10%
Carbon-14	14.0032420 u	~0%
Oxygen-16	15.9949146 u	99.762%
Oxygen-17	16.9991312 u	0.038%
Oxygen-18	17.9991603 u	0.200%

* 1 $u = 1.6605402 \times 10^{-27}$ kg

What those abundance percentages mean can be illustrated by referring to carbon. Approximately 99 out of every 100 carbon atoms you would find are carbon-12. On occasion, an atom of carbon-13 would be found, one per 100 carbons.

Since natural carbon is a mixture of mass 12 and mass 13 isotopes in the percentages indicated, the *average* atomic mass* would be approximately $0.99 \times (12) + 0.01 \times (13.00) = 12.01$. In the case of carbon, this differs from 12 by a very slight amount. But for other elements, the natural isotopic mixtures can vary significantly from whole numbers. Also, much less obvious now, the excellent precision with which the abundances and masses can be determined turns out to be an extraordinarily useful target for analysis in carbon dioxide studies – its legacy – as we will see in later chapters.

The isotope carbon-14 is radioactive and is naturally occurring to the extent of 1.2×10^{-10}% in the atmosphere. It was discovered in 1934 by American physicist Franz Kurie of Yale University who bombarded nitrogen with neutrons which themselves had been discovered in 1932. Carbon-14 is frequently, but ambiguously, referred to as *radiocarbon*.

These are not the only known isotopes of carbon and oxygen. A dozen more isotopes of each have been discovered. All are radioactive, most with very short lifetimes. The identity and lifetimes of these curiosities are indicated in the table.

Other Known Carbon Isotopes	Average Lifetime	Other Known Oxygen Isotopes	Average Lifetime
C-8	?	O-12	?
C-9	0.18 seconds	O-13	0.01 seconds
C-10	28 seconds	O-14	1.7 minutes
C-11	29 minutes	O-15	2.9 minutes
C-15	3.5 seconds	O-19	39 seconds
C-16	1.1 seconds	O-20	19 seconds
C-17	0.28 seconds	O-21	4.9 seconds
C-18	0.14 seconds	O-22	3.2 seconds
C-19	0.07 seconds	O-23	0.10 seconds
C-20	0.02 seconds	O-24	0.07 seconds
C-21	Less than 30 nanoseconds	O-25	?
C-22	0.006 seconds	O-26	4.5 picoseconds
		O-27	Less than 260 nanoseconds
		O-28	Less than 100 nanoseconds

* Some literature refers to atomic mass as atomic weight. Strictly speaking, this is incorrect because weight is something you determine by weighing and depends on gravity. On a roller coaster ride, you can be briefly weightless. On the planet Jupiter, you would be very heavy. But your mass would remain the same. Different gravitational attraction would result in different weights for the same object. However, in discussing the isotopes and elements as we have done briefly here, everything is done relative to carbon-12 and so the values are referred to as *relative* atomic weights.

In the table, you can see some of the lifetimes have not been determined yet. Expectations are that a few more isotopes are yet to be discovered.

C-11 and O-15 are both used in sophisticated medical positron emission tomography (PET) scans because their radiations are easily measured and the locations of the radiation-emitting atoms can be pinpointed with good geometrical resolution.

ATOMS

An atom, more or less, is 100,000 times as large as its bare nucleus. Put in a more pictorial perspective, if the typical atomic nucleus were as large as the period (".") at the end of this sentence, then the entire atom with its collection of electrons would be the size of a football arena.

Electrons are spread around the nearly point-like central positively charged nucleus. Detailed electron behavior is beyond the scope of our discussions and is also quite esoteric in nature. All we need concern ourselves with is that the electrons are assigned to shells whose average distance from the nucleus gets larger and larger as each new shell first receives electrons.

The assignment of electrons, at least for the first 20 elements with atomic numbers (proton count) ranging from 1 to 20 is that

- The first two electrons go in the first shell.
- The next eight electrons are assigned to a second shell.
- The next eight electrons are assigned to a third shell.
- The next electrons start a fourth shell.

When a shell has its full complement of electrons, it is referred to as a filled shell. The significance of a filled shell is that it has unusual stability, that is, resistance to change. Below is a table of the first 20 elements arranged in order of increasing atomic number starting with hydrogen (element 1) in the upper left and reading left-to-right to calcium (element 20). The arrangement shown here allows a visualization of the repeating physical and chemical properties of the elements associated with the number of electrons in the last shell being filled. For example, the closed shell electron arrangements associated with total electron numbers of 2, 2 + 8 and 2 + 8 + 8 correspond to the eighth column or family of elements, called noble gases, all of which are very unreactive chemically. In contrast, the seventh column elements, with electron arrangements of 2 + 7 and 2 + 8 + 7, are called the halogen elements or halides. For elements in the first column, on the left, the electrons are configured as 1, 2 + 1, 2 + 8 + 1, and 2 + 8 + 8 + 1. Except for the lightest of these, hydrogen, the group is called the alkali elements, having just a single electron in the last shell.

H hydrogen							He helium
Li lithium	Be beryllium	B boron	C carbon	N nitrogen	O oxygen	F fluorine	Ne neon
Na sodium	Mg magnesium	Al aluminum	Si silicon	P phosphorus	S sulfur	Cl chlorine	Ar argon
K potassium	Ca calcium						

This is an abbreviated version of the Periodic Table first articulated nearly completely by the Russian chemist, Dimitri Mendeleev, in 1869. Although understanding the *causes* of the periodicity in properties had to await the discovery of the electron, of x-rays, and of radioactivity, we observe here that the properties, including the reactions of the elements, are influenced almost entirely by the number of electrons in their last shell, the so-called *valence* (from Latin *valentia* = capacity) electrons. For the first column (or "family" or "group"), there is one valence electron; for the seventh column there are seven valence electrons; for the eighth column, there are eight valence electrons (except for helium).

COMPOUNDS AND MOLECULES

Atoms, positively charged nuclei plus their surrounding electrons, are able to lose or gain electrons, thereby becoming *ions*. Atoms that lose electrons become positively charged ions. Atoms that gain electrons become negatively charged ions. With their opposite charges, a positively charged ion and a negatively charged ion attract each other and can form a combination called a *compound*. Table salt, for example, is basically sodium which has lost an electron and chlorine which has gained an electron. That is, Na^+ and Cl^-. Persons on low-salt diets sometimes use a salt alternative that is K^+ and Cl^-: potassium chloride. Both sodium and potassium (as well as the heavier alkali elements) very readily tend to lose their one valence electron in forming compounds. The reciprocal of this is that chlorine and the halogens have a tendency to gain an electron in forming compounds. A simplified explanation of this is that in doing so, both the alkali elements (column one) and the halogen elements (column seven) are left with closed shell electron arrangements which we noted earlier are extra stable. Since much of chemistry involves combinations of the elements depicted above where the closed shells have eight electrons, it is common to see reference made to an *octet rule* as an indication that changes seem to lead to the closed shell structure. Although chemistry, the science of such transformations, is actually much more complicated than this, the octet rule can be used to model an enormous amount of what transpires.

A compound is a combination of atoms of at least two different elements. It is one kind of *molecule*. A molecule is a combination of at least two atoms, which need not necessarily be different. Carbon is unique in the variety of ways in which it can combine with other carbons and/or other elements. The majority of known compounds contain carbon. A very simple class of carbon compounds is the hydrocarbons, consisting of just the element carbon combined with the element hydrogen. Natural gas encompasses the C_1 through C_4 hydrocarbons (in which the subscript indicates the number of carbons with the number of hydrogens left unspecified); gasoline encompasses C_6 through C_{10}; diesel fuels C_{14} through C_{30}; and lubricating oils C_{26} through C_{40}. There is an incredible number of possible molecules made from carbon and hydrogen combinations. Just the C_{30} hydrocarbons with the maximum number of hydrogens – 62 it turns out – all having the formula $C_{30}H_{62}$, comprise over 4 billion different possibilities.

Our attention will focus on the compound carbon dioxide in subsequent chapters. For now, very briefly, the carbon dioxide molecule has one carbon atom combined with two oxygen atoms. Carbon dioxide has an analog that is also a very common terrestrial substance and worth mentioning. In the abbreviated Periodic Table, you could note that just below carbon in the fourth column, as part of its chemical "family," is the element silicon, symbol "Si." This element is the basis of the huge semiconductor industry, one of the commercial centers of which is branded "Silicon Valley" in California. The combination of silicon with two oxygen atoms is the counterpart to carbon dioxide and is called silicon dioxide. But even though many properties of silicon and carbon are similar, the differences are sufficient such that SiO_2 is vastly distinct from CO_2. Silicon dioxide is the essential component of quartz and the major constituent of "sand." Silicon is the second most abundant element in Earth's crust. Its 26% abundance is exceeded only by that of oxygen.

For many years, it was thought that the only combinations of atoms possible were those in which positive and negatively charged species were involved. Amadeo Avogadro adopted the seventeenth century term "molecule" in 1811 to indicate combinations of identical species. He did this to explain some very simple experiments involving gases and the simple relationship among the volumes of combining gases and produced gases discovered by the Frenchman Gay-Lussac. It was Avogadro who proposed that the elemental form of oxygen consisted of two atoms in a molecule that we would now write as O_2. Hydrogen gas was proposed to be H_2. Avogadro's hypothesis was resoundingly rejected by scientists at the time because there was no way to understand what would hold two oxygen atoms together since they were both neutral: uncharged. And there was also the unanswered question as to why just two atoms were involved. Why not three, or four, or more? These were extremely reasonable questions at the time, but were not evidence that the hypothesis was incorrect. The questions merely illustrated that the phenomenon of bonding in chemistry was not understood in the early 1800s. Patterns and trends can be misleading.

2 Early Earth and Our Solar System

The thinker makes a great mistake when he asks after cause and effect. They both together make up the indivisible phenomenon.

GOETHE

Before commencing the detailed drama of carbon dioxide, we must continue to set the stage: our planet Earth. A quick look at our solar system and some ideas about it will serve this purpose well. Omitted from the discussion is the limited degree of confidence associated with some of the descriptions. This chapter should be peppered with "maybe," "perhaps," "it is thought," and other qualifying expressions. The concepts are all reasonable, though, and science has continuously adjusted its viewpoint as more and more sophisticated studies are done.

GESTATION: FORMING PLANETS WITH ATMOSPHERES

Based on analysis of meteorites – solid debris from outer space – many astronomers are convinced that a supernova explosion preceded the formation of our sun when the nearly 14-billion-year-old universe was two-thirds of its present size. The birth of the solar system is dated to just under 5 billion years ago. Gravitational condensation of a massive cloud of dust and gas whose composition was thought to be like that of the sun formed around the solar core. From this contraction, we got the sun and the planetary disk. The first person known to have suggested that a nebular origin led to the solar system was the Scot-descended German philosopher, Immanuel Kant. Following his telescope observation of spiral nebulae, Kant wrote in detail his theory of the origin of the sun and the planets in *Universal Natural History and Theory of Heaven* in 1755. But the publisher went bankrupt, leaving the book essentially unknown for decades.

As the solar nebula contracted and heated up, the volatile elements were driven to the outer portion of the planetary disc that was taking shape. The inner planets that formed close to the massive sun were correspondingly enriched in heavier elements. Recently, evidence using the slow radioactive decay of ^{238}U (uranium is element 92) into Pb (lead, element 82) implies that Earth as a planet may have accreted as early as 30 million years after the formation of the solar nebula. And even more recent studies suggest this might have taken as little as 3 million years to be accomplished. Metals would sink to form the core and silicates rise to form the crust rapidly, taking perhaps 1 million years.* The initial atmospheres of the inner planets, including Earth, should have been mutually similar. The outer planets were endowed with light elements. For the inner planets, volatile substances, those that are gases at warm to moderately high temperatures, were not incorporated into planet formation for the most part. Exceptions would be those substances, oxygen and carbon dioxide for instance, that could chemically react to form nonvolatile solids such as oxides and carbonates. Volatility, on the other hand, accounts for the rarity on Earth of the noble gases, the eighth group in the Periodic Table containing helium, neon, argon and more.

Yet, in seeming contradiction, helium is modestly available (for balloons…and research) and argon is the third most abundant gas in our atmosphere now. Helium and argon are two anomalies whose presence is well known to be due to radioactive decay. In the case of helium, its occurrence is a consequence of the radioactive decay in Earth's core of uranium- and thorium-containing

---

* T. S. Kruijer et al., "Protracted core formation and rapid accretion of protoplanets," *Science* 344, 1150–1154 (2014).

substances, both of which are *alpha-particle* emitters. The alpha-particle is identical to a helium atom which lacks its two electrons at emission but quickly picks them up. As far as argon is concerned, its origin is in the potassium (K) in Earth's crust. Potassium, the 19th element, consists naturally of three isotopes, mostly ^{39}K, but with about 6.7% ^{41}K and 0.012% ^{40}K. The last is radioactive and decays most of the time to ^{40}Ar, a stable isotope of the 18th element. Almost all the argon in the atmosphere is ^{40}Ar and owes its abundance to the decay of ^{40}K at a rate determined by the ^{40}K half-life of 1.4 billion years (or average lifetime of 2 billion years). The amount of ^{36}Ar, another stable isotope of argon, is a million times *more* abundant than ^{40}Ar in the sun than in Earth's atmosphere. That anomaly is consistent with the total loss of all volatile primal argon during the initial formation of our planet. Similarly, the Mars Rover in 2013 determined that argon in the Martian atmosphere is 99.92% ^{40}Ar.*

Argon isotope abundance is part of the evidence that Earth did not retain its primary (original) atmosphere, for if it had, the relative abundances of the various argon isotopes would have been much more solar-like. The comparison below shows solar system abundances of some elements alongside terrestrial abundances, all adjusted in comparison to an arbitrary benchmark of "10,000" for silicon, a very abundant element in both domains.

Element	Solar System Abundance Relative to Si = 10,000	Terrestrial Abundance Relative to Si = 10,000	% Retained Terrestrially Relative to Si
H	350,000,000	84	0.000024
He	35,000,000	0.00000035	0.000000000001
C	80,000	71	0.09
N	160,000	21	0.013
Ne	50,000	0.0000012	0.0000000024
Na	462	460	≈100
Al	882	940	≈100
Si	10,000	10,000	

The difference in relative abundances for the volatile elements is striking, especially for the noble gases helium (He) and neon (Ne). For all intents and purposes, these form no stable and no nonvolatile compounds. Their minor presence may be accounted for by recognizing that trace quantities do adhere to surfaces like those of dust particles. Compared to the volatiles, the last elements tabulated above manifest the opposite behavior. Being extremely nonvolatile when in compounds, as is usually their situation, they are referred to as *refractory* and, like most of the other refractory substances (not shown), have relative abundances that likely mirror those in the primary nebular cloud from which the solar system arose.

EARLY ATMOSPHERIC ACTIVITY

There is little surviving evidence of the early accretion scenario, that is, the first half-billion years. But there are models. One model has a "blowoff" commencing about 50 million years after the sun finished contracting. The "blowoff" is a rapid, hydrodynamic outflow of mostly hydrogen that, like a swift wind, carries along other, heavier gases. In 2003, the blowoff process was observed on an extrasolar planet – Osiris – in the constellation Pegasus some 150 light years away. The amount of

---

* P. R. Mahaffy et al., "Abundance and isotopic composition of gases in the Martian atmosphere from the Curiosity rover," *Science* 341, 263–266 (2013).

hydrogen required for the blowoff model amounts to some 88 oceans worth of hydrogen, that is, from the hydrogen in that much water. This seems like a huge amount, but the large, outer planets of our solar system are known to still contain ~50% ice. Yet since hydrogen is a volatile gas, it should no longer have been present by the time Earth's accretion concluded. Nevertheless, the hydrogen could have arisen externally, from acquired gas or even from water, the latter being broken apart by energetic light hitting the atmosphere in a well-known process called photolysis. Moreover, frequent meteor and comet impacts on the cooled but young planet could easily lead to accrual of materials just like those that had been previously purged. Photographs of Earth's atmosphere taken from the moon and filtered through wavelengths ("colors," in a sense) specific to hydrogen atoms unequivocally show the presence of the gas hydrogen.

Helium is exceptionally low in abundance in the atmosphere despite its constant production by alpha-decaying radioactive isotopes in Earth's crust and mantle. The light gas escapes from the atmosphere because of the speed with which it is moving when at the temperature of the surroundings. But this obvious mechanism of thermal escape of light gases from a hot early Earth is no longer the favored picture for the loss of volatiles (although it could still be invoked for planetesimals – small protoplanets – as they accrete). It appears, though, that there is as yet no way to distinguish these volatile loss mechanisms from an alternative picture in which there was incomplete condensation of the initial nebular gas during planet formation.*

ATMOSPHERIC INFLUENCES DURING EARTH'S INFANCY

In the absence of any atmosphere, overall temperature balance on a planet is due mostly to input (radiation from the sun, also known as *insolation*) minus reflection (*albedo*†). Using Mars's measured albedo and assuming that solar luminosity has been constant (so as to simplify the estimation), Earth's primitive surface temperature is calculated to have been about 260 kelvin (−13°C or 9°F). Reflected radiation portions can exceed 80% from thick clouds or from uncompacted snow. Water bodies have albedos around 10% and vegetated areas less than 20%. An early scene, startlingly, would be one of an essentially airless Earth. And frigid.

A clever way of estimating what portion of the present atmosphere might be remnants of a primary atmosphere is to use the abundance of the gas neon as a yardstick. That is, use the current, relatively low amount of neon in the air as representative of the original atmosphere still around, that is, as a proxy. Nitrogen and oxygen gas levels relative to neon should be similarly low. For example, the previous table shows nitrogen in the sun to be twice the concentration of neon in the sun. But today, nitrogen is millions of times more abundant in air than neon. The reasonable interpretation is that only an extremely small fraction of the current atmosphere, mostly nitrogen, is original. And that comparison value is a significant overestimate because the neon now in the atmosphere was probably trapped in Earth's interior eons ago rather than being in the original air itself. Neon is a problematic proxy and turns out not to help in confidently pinning down ancient air contribution.

The history of the planet's physical and chemical evolution is fascinating on its own merits and constantly being studied with improving technology. Our very brief review here does it no justice.

Many scientists believe that much of the water of the oceans and of the gases in the early atmosphere was deposited here by comets and/or meteors and/or asteroids. This hypothesis is now being scrutinized through flybys of comets exploring relevant isotope ratios in comparison to those on Earth. The accretion of carbon for carbon dioxide is somewhat of a puzzle though. This is because the simplest carbon compounds such as carbon dioxide, carbon monoxide, and methane (CH_4) won't exist as ices at times of accretion of the inner (warm) planets. They are volatile and would be lost as vapors. However, heavier hydrocarbons are a possible source of carbon accretion and could

---

* A. N. Halliday, "The origin and earliest history of the earth," *Treatise on Geochemistry*, 2nd edition, 1, 149–211 (2014).
† From Latin, *albus*, white.

subsequently react chemically with water to produce carbon dioxide and hydrogen, the latter then most likely escaping the atmosphere.

Truly, early planet structures were presumably high temperature, molten bodies. The high temperatures were due to energy released as the planetary masses collapsed under their own gravity; to energy deposited from bombardments by massive meteors abundant during the early age of the solar system; and to radioactive heating greater than four times today's rate, due to the much greater abundance of not-yet-decayed but short-lived unstable elements. (Among these is a hafnium isotope, ^{182}Hf, of element 72 with a 9-million-year half-life and which will be discussed below.) Evidence of the meteor impacts is easily recognized by looking at the crater-scarred surface of the moon and the planet Mercury. At such high temperatures, not only was outgassing probable, but the force of the exiting vapors' blowouts could carry away lingering atmospheric gases much like a strong wind can carry pollution and light debris along with it. Gravitational collapse as a source of heat likely would have lasted only for some millions of years. As our planet matured, decreased frequency of impacts and slowing gravitational collapse allowed cooling to commence. Evidence based on noble gas isotope data implies that the early atmosphere's mass was much greater than it has been in recent eons. Such blanketing would have allowed surface temperatures to be several thousand degrees, enabling the existence of oceans of magma.*

A stable crust of less dense solids such as silicates would form as planet cooled. Some elements such as hafnium are more soluble in the molten silicates than in the melted core whereas some elements, such as number 73, tungsten (W) are more soluble in the core. Separation between these two would then occur. Studies of crustal ^{182}W, the isotopic product of ^{182}Hf decay,† indicate core formation occurred on the order of only 1 million years after planetary accretion, followed by silicates floating up to form the surface crust. Thermal energy release by volcanic activity would be the major cooling phenomenon. Volcanic outgassing gave rise to what we could call our secondary atmosphere: methane, nitrogen, hydrogen, ammonia, water vapor, carbon monoxide, and carbon dioxide. The sequence of volcanic gas abundance is believed to be that just listed, decreasing from methane. Ultraviolet light (UV), with energy high enough to disrupt chemical bonds, would arrive from the sun. The UV would break down some of the hydrogen-containing molecules, releasing hydrogen atoms. Owing to their lightness, H atoms would escape the gravitational pull of the planet and leave the atmosphere, though the intense bombardment by extraterrestrial bodies during these early (millions of) years was an auxiliary route to replenished volatile substances. In a relatively short time, the atmospheric composition would change substantially, becoming mostly carbon dioxide and nitrogen. This is also the case for the planets Venus and Mars. And still is, for them, as indicated below with their distances from the sun in parentheses; however, it is no longer so for our home planet.

	Venus (67 Million Miles)	Earth (93 Million Miles)	Mars (141 Million Miles)‡
Surface Temperature	745 K	280 K	225 K
Atmospheric Pressure	90 atm	1 atm	0.01 atm
CO_2	96.5%	0.035%	96.0%
N_2	3.5%	78%	1.89%
O_2	0.003%	21%	0.145%
Ar	0.003%	0.9%	1.93%

* A. N. Halliday, op. cit.
† T. S. Kruijer et al., *Science* 344, 1150 (2014).
‡ P. R. Mahaffy et al., op. cit.

Although hydrogen escapes Earth's atmosphere, it can be regenerated by chemical reactions. The reaction of water vapor with light of sufficiently short wavelength and therefore high enough energy is still occurring today as evidenced by x-ray photographs of Earth taken in 1972 by Apollo 16 from the moon which filtered the exposure so that only emissions from atomic hydrogen appear.

SIBLING PLANETS

Early Venus probably had as much water vapor and/or liquid water as Earth. But this water was lost due to a "runaway greenhouse effect." Being closer to the sun, Venus receives about twice the solar heat that Earth receives. The humidity must have been huge due to evaporation of any liquid water into the atmosphere. Water vapor, like carbon dioxide, is an effective greenhouse gas. With vast amounts of H_2O in the atmosphere, decomposition of water by ultraviolet light would release hydrogen as atoms which would escape the atmosphere or combine with other atoms, perhaps with another hydrogen to form molecular H_2, which is still light enough to escape. The present temperature on Venus is very high at 745 kelvin, as the above table shows. Obviously, in such heat there would not be rainfall,* nor would there be photosynthetic life. Since rain provides a means for weathering rocks (Chapter 16), and photosynthesis (Chapter 14) consumes CO_2, both very effective processes on Earth for removal of carbon dioxide from the atmosphere, Venus's CO_2 has not been reduced from its high primordial level.

Mars, being further from the sun, is at a lower surface temperature than Venus (and Earth). Condensation of water vapor occurs, but as ice and snow. It is estimated that Mars originally had perhaps 30 atmospheres worth of water condensed as evidenced (but not proven) by stream beds that are observed and from geological knowledge of how such remnants must have been formed. The average temperature was not higher than 220 kelvin (−50°C or −58°F). Both warmer and colder regions would exist too. When the carbon dioxide pressure got high enough, CO_2 ice could form in the colder, polar regions. Although the relative abundance of carbon dioxide on Mars is high at 95%, the total amount is low since the air pressure, a measure of the total amount of atmosphere, is 1% of the atmospheric pressure on Earth. Calculations suggest that the formation of carbonate minerals (Chapter 12) has served to store as much Martian carbon dioxide as is the case on Earth, although not for all the same reasons.

The temperature *versus* vapor pressure diagram here for water emphasizes the differences among the three planets. The solid curved line rising from the lower left to the upper right delineates vapor-condensation equilibrium where vapor and liquid or solid co-exist. (Liquid and solid are the "condensed" phases.) It indicates the conditions of temperature and pressure at which water vapor condenses to either liquid water (rain) or solid water (snow, ice). Where that curve hits the right edge of the figure frame corresponding to a water vapor pressure of one atmosphere on the bottom axis is the normal boiling point of water, 373 K (100°C or 212°F) where the liquid and gas are both present. The horizontal line separating liquid from ice on the right is at the normal freezing temperature: 273 kelvin (0°C or 32°F). It intersects the curved solid line at the triple point where vapor, liquid, and solid all naturally can co-exist. That is the only condition under which the three can be present at equilibrium. Those solid lines are thermodynamically fixed properties of water and independent of environment.

---

* The surface temperature of Venus may have been so high (647 K, the "critical point") that the liquid form of water thermodynamically did not exist.

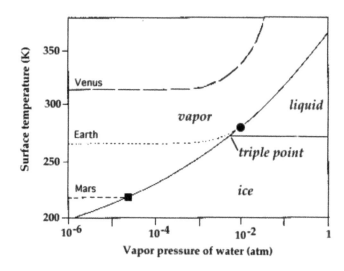

Starting on the left of the figure near the bottom, the flat dashed line indicates the surface temperature of Mars as 220 kelvin (–50°C). If water vapor pressure at that temperature grew from near zero (10^{-6} atm on the bottom axis) to the square data point at about 0.00002 atmospheres, any additional water vapor from yet higher pressure moving to the right on the figure would condense out as ice (or snow), since the vapor/ice equilibrium curve would be crossed (at the black square). Where the dashed line at flat pressure would be extended, no gaseous water would exist on the too-cold Martian surface. In contrast, Venus starts out much warmer, at about 315 kelvin. Following the upper, long-dashed line, as the vapor pressure – the amount of water in the atmosphere – increases, the greenhouse effect begins to set in above 0.001 atmospheres. (That value is for water vapor pressure, not total atmospheric pressure.) The long-dashed line abruptly curves upward showing that the surface temperature on Venus would rise rapidly as the water vapor content of the atmosphere continued to grow, powering the greenhouse warming. If additional water vapor accumulates, the vapor would never cross either the vapor/ice equilibrium curve nor the vapor/liquid equilibrium curve. Venus would have neither ice nor liquid water.

Earth sits delicately between the Venus/Mars extremes. With increasing amounts of water vapor, the temperature would remain constant until the greenhouse effect emerges above a few thousandths of an atmosphere of water vapor pressure, indicated by the dotted line in the figure. The temperature rises as more water vapor is added, but not as rapidly as the condensation equilibrium curve (solid curve) is rising with vapor pressure. At just above the "triple point" for water (where vapor, liquid, and ice can all three co-exist in equilibrium), the amount of water vapor in the atmosphere hits the maximum possible at that temperature – saturates – and condenses into liquid at the filled circle in the figure at 280 kelvin (7°C or 45°F) – rain.

THE TODDLER EARTH

Earth's early atmospheric chemicals were continuously resupplied by accretion from then-abundant comets and meteors. Water in the atmosphere is represented at this ancient stage by the left portion of the horizontal dotted Earth line in the figure. As the water vapor pressure in Earth's atmosphere increases from about 0.01% of the present pressure, greenhouse warming begins and the temperature rises above that of the ice–liquid transition temperature. Liquid condenses from the atmospheric vapor, that is, it rains. Additional water accretion from extraterrestrial sources enables more and more rain, leading to river formation and oceans, but not to more atmospheric water vapor pressure. A fixed water vapor pressure is a fixed amount, not a fixed percentage: you don't get absolute humidity above 100%.

Earth's atmosphere, notable by comparison to the other two planets, is different…fortunately. That Earth's temperature fell in the range of liquid water rather than vapor (Venus) or ice (Mars) allowed for an additional pathway to redesign Earth's surface. We'll get to that later. However, it is well recognized that the sun was roughly a quarter less luminous during these early ages; enough so, that if this were the sole consideration, the planet should have been covered with ice for 2 billion years, nullifying the existence of liquid water and even life for which there is evidence after the first billion years. That expectation is countered by the known effects of heavy concentrations of greenhouse gases in the atmosphere. Many uncertainties pertain to this early era and interpretations are very model dependent.* Among the visual evidence for a wet environment are mud cracks and ripple marks in sediments indicating the presence of liquid water, confidently dated by radioactivity techniques to layers older than 3 billion years. Earth's atmosphere has constrained temperatures to be moderately warm at a very roughly constant level, except for arguable periods in antiquity when extreme ice ages pertained. A fascinating state of affairs in its own right, ice-encrusted Earth and the role of carbon dioxide will be discussed in Chapter 17.

ADOLESCENT EARTH

Besides atmospheric changes, there are alterations in the solid structures of the planet's surface. Asteroid impacts during early stages of the planet's history are presumed to have been of such intensity and size as to evaporate oceans to 3 km depths or more and to sterilize the surface repeatedly. Much of Earth's earliest surviving crust dates back 3.8 billion years when the heavy asteroid bombardment likely faded (based on studies of lunar craters). Yet, the mineral known as *zircon*, highly resistant to weathering, transport, re-deposition, and erosion, has been used to establish a record longevity for the oldest crustal rock at 4.374 billion years.† Because of such initially hostile environmental conditions, the first 700 million years of Earth's existence are referred to as the *Hadean* Period (after Hades). Some estimates suggest that during the Hadean, as many as 15 objects at least 100 miles wide struck Earth. Even moon-sized objects probably swept close by, driving major changes in the atmosphere. Complex life forms seem only to have arisen permanently several hundred thousand years later.‡

Geographic variation usually occurs extremely slowly, roughly measuring an average of a millimeter per year: up, down, sideways, whatever. Of course, there are exceptions at extremes. Millions of years of moving and colliding geologic plates raise mountain ranges. Rain, wind, and chemistry reshape and recycle the mountains into clay, silt, and dissolved substances that find their way eventually into the oceans, often depositing onto ocean bottoms. Earth may be somewhat unique in that its solid crust is continuously being destroyed, churned and regurgitated: a massive natural recycling scheme. Despite the sluggishness with which these processes typically transpire, there are occasions in which events take place with explosive rapidity. Besides conspicuous volcanic activity, there are asteroids, meteors, and comets that smash into Earth at thousands of miles per hour. Impacts can melt rock – even vaporize it. Shockwaves can shatter crystal structures far away and can demagnetize some rocks. Among the best known of these phenomenal episodes involves Arizona's Meteor Crater, blown out perhaps 50,000 years ago. The crater is 1.2 km in diameter and 200 m deep. Agreement is that the perpetrator of this landmark was an iron–nickel meteorite perhaps 45 m in diameter, half the length of a soccer field. That is roughly the size of the small whitish area in the picture below at the center of the crater bottom.

---

* G. Feulner, "The faint young sun problem," *Rev. Geophys.* 50, RG2006 (2012).
† J. Wraltey et al., *Nat. Geosci.* 7, 219 (2014).
‡ There is a report on a "potentially biogenic carbon" in a 4.10 ± 0.01 billion-year-old zircon from Western Australia published by E. A. Bell, P. Boehnke, T. M. Harrison, and W. L. Mao in *Proc. Natl. Acad. Sci.* 112, 14518 (2015).

EARTH'S PUBERTY

Only a half-century ago, scientific experiments suggested the origin of life, some 3.5–4 billion years ago, could be ascribed to reactions involving methane and ammonia in the atmosphere. The latter would be what's called a chemically "reducing" atmosphere, the opposite of an oxidizing atmosphere that we now have. Electric discharge experiments (lightning in the laboratory) by Stanley Miller, working with Harold Urey in the 1950s, illustrated the formation of complex mixtures of organic substances, some of which were amino acids, arguably portending the possibility of life. Accumulated organic molecules in the early ocean were referred to as the primordial soup. In the early nineteenth century, Russian Alexander Oparin, followed shortly thereafter by Englishman J. B. S. Haldane, proposed such a scheme for the origin of life. They recognized also that oxygen could not have been present because it would have chemically detoured the production of many of the soup's ingredients. But such an atmospheric recipe was not supported by further studies in the 1960s.

Very recently, unpublished results from Miller's laboratory showed that he also used an apparatus that mimicked a volcanic eruption accompanied by lightning. Miller also included hydrogen sulfide (H_2S) in the gas mixture. The new analyses revealed that over 40 different amino acids and amines were generated, suggesting volcanic island arc systems could have provided an environment for some of the processes thought to be involved in chemical evolution and the origin of life.

On the other hand, it was realized that Miller's production of amino acids was probably not the critical initiation step for the origin of life because amino acids and amines are not self-replicating, a key characteristic of living systems.

Also, recently, it was hypothesized and soon more widely accepted from indirect evidence that the early atmospheric methane and ammonia gas abundances would have been scarce. Carbon dioxide would have been plentiful. At first glance, the production of organic substances is much less apt to occur in such an atmosphere – bad news. However, a group of planetary scientists led by Feng Tian from Colorado proposed in 2005 that, contrary to the conventional picture, hydrogen would have been a major constituent of the young Earth atmosphere, comprising possibly as much as 30% of the total gas composition.* In this vision, the missing ingredient for synthesis of organic matter by electrical discharges in the atmosphere is re-established. Contradicting earlier mentioned considerations of the atmosphere's composition, Tian and co-workers argued that hydrogen supplied by volcanic outgassing was not as efficiently able to escape the planet as previously predicted.

There do exist competing scientific ideas for life's launching, one divorced from the atmosphere. In the ancient oceans, sulfate (the highly oxidized form of sulfur, element 16), would have been prevalent. The abundance of sulfurous emissions from submarine hydrothermal vents would be ideal

---

* F. Tian, O. B. Toon, A. A. Pavlov, and H. DeSterck, "A hydrogen-rich early earth atmosphere," *Science* 308, 1014–1017 (2005).

for organic synthetic reactions leading to complex structures, potentially with biological relevance. Such a model is now seriously considered as an alternative to the atmospheric source concept.

And in 1970, analysis of the 100 kg *Murchison meteorite* revealed the presence of extraterrestrial amino acids suggesting their syntheses occurred during the early history of the solar system. The Murchison meteorite was an important finding since its 1969 descent and touchdown outside Melbourne, Australia, was widely observed and the meteorite was recovered and analyzed very quickly, minimizing the chance of contamination.

Ultimately, all the different pathways may have contributed in different degrees to the characteristics of young Earth.

COMMENCEMENT

All the processes occurring over Earth's history continue to shape the carbon dioxide narrative: volcanism, erosion, photosynthesis, precipitation, subduction, weathering, combustion, decay, and evaporation. Let the legacy story commence.

3 Discovery

A man should look for what is, and not for what he thinks should be.

EINSTEIN

It is ironic that something as ubiquitous as carbon dioxide, a substance that certainly has been around at least since early Earth, wasn't really noticed until a few hundred years ago.

VAN HELMONT

The Belgian chemist and physician Jan Baptista van Helmont was likely the first to recognize that what we call air was not the only gaseous substance in existence. At the end of the sixteenth century and for yet another 150 years, the term *air* meant gas. Van Helmont, who lived from 1580 to 1644, recognized that the substance given off by burning charcoal was identical to that yielded by the fermentation of grapes. Van Helmont invented the word "gas" to describe the *spiritus sylvestre* (spirit of wood) emanating from the burning charcoal. The word derives from the Greek word "chaos" (χάος), although alternative, but erroneous, etymologies suggest the word derives from the Dutch word *geest* akin to the German *geist* meaning "spirit" or "ghost." The Oxford English Dictionary has a 1662 quoted translation of Helmont's work that reads:

> Because the water which is brought into a vapour by cold, is of another condition, than a vapour raised by heat: therefore...for want of a name, I have called that vapour, Gas, being not far severed from the Chaos of the Auntients. Gas is a far more subtile or fine thing than a vapour, mist, or distilled Oylinesses, although as yet, it be many times thicker than Air. But Gas it self, materially taken, is water as yet masked with the Ferment of composed Bodies.

Jan Baptista van Helmont

The last sentence in the quote is certainly strange. Van Helmont regarded a gas to be contained in all bodies as an ultra-rarefied condition of water. Van Helmont was a believer in water being a basic element of the universe, as were the ancient Greeks and many thereafter. Quantitative experiments were relatively new in his time, but he recognized their utility, even though many philosophers frowned on the lowly concept of actually making measurements. His most noteworthy quantitative evaluation was growing a tree in a weighed amount of soil, nurturing it with weighed quantities of water and noting that the 164-pound-increased weight of the tree, taking into account fallen leaves, did not come from the soil but rather, in his view, from the added water. Carbon dioxide, of which van Helmont himself had become aware, was overlooked as a required nutrient.

HALES

A half-century after van Helmont, the English botanist and physiologist Stephen Hales advanced methods for studying different gases by inventing a device to collect gases over water. Hales originally studied theology at Cambridge but was active in science on the side. Among his accomplishments were measurements on the growth of plants and the pressure of sap. Hales, too, was the first to measure blood pressure. Also, he made a significant contribution to public health with studies on the beneficial effects of fresh air ventilation. Hales became aware that some air contributed to the growth of plants, thus modifying van Helmont's misinterpreted weight gain experiments from the previous century. His 376-page work was published as *Vegetable Staticks* in 1727. Staticks was the science of weights and measures at the time.

Stephen Hales

BLACK

Joseph Black, one of 15 children, was born in Bordeaux, France, on April 16, 1728. His father was a wine merchant of Scottish descent and his mother was from a Scottish wine trade family. He was sent by his parents first to Belfast and subsequently to Glasgow to study "arts," that is, Latin and Greek. A few years later, he was convinced to take up a more marketable venture and chose medicine. At Glasgow University, his professor of medicine had just established lectures in the field of chemistry, at that time regarded as a sub-discipline of medicine. Black's mentor was noted for encouraging students to engage in independent experiments. The medical professor, William Cullen, hired the 20-year-old Black as an assistant in the laboratory. Subsequently, Black pursued further medical studies in Edinburgh after which he returned to Glasgow as a Lecturer in chemistry. During the early Glasgow years, he began research for his MD thesis on the chemistry of *magnesia alba*, which means "white magnesia." This is a form of hydrated magnesium carbonate sometimes called mild magnesian earth. Named after the ancient city in Asia Minor, Magnesia (in western Turkey, southwest of Smyrna), it has the formula $4MgCO_3 \cdot Mg(OH)_2 \cdot 5H_2O$. Suspensions of magnesia alba are still used as milk of magnesia. Black's 1754 dissertation was titled *"De humore acido a cibis orto, et magnesia alba"* (On the Acid Humour Arising from Food, and Magnesia Alba). The medical dissertation, written in Latin as was the tradition at the time, dealt mostly with the use of magnesia as an antacid. But it also included the discovery of carbon dioxide. An English translation was not made available for over 150 years despite the importance his discovery of carbon dioxide had in the field of chemistry.

> I have made no few experiments: many of them were new, and some even worthy enough of record: I therefore thought that it would not be unpleasing to those of my fellow students who are fond of chemical philosophy if I were to publish the more remarkable of them: and so I hope that they will be favourably received by such. For the result, whatever it be, of a new experiment is not to be neglected; since the foundation of chemical science, so useful to medicine, and yet still so very imperfect, rests on experiments alone.

The experiments, which led to the discovery of what Joseph Black termed "fixed air," involved for the first time very careful weight measurements of *magnesia alba* as it was heated (to release carbon dioxide as the result of thermal decomposition). He followed the reactions of the products with acids or alkalis. Black observed that magnesia "is dissolved quickly and completely by vitriolic acid, and in their union a great abundance of air is expelled from them." Vitriolic acid is the old name for sulfuric acid. The same occurred with acid of nitre (nitric acid) and distilled vinegar (acetic acid). Remember, at the time, the term "air" meant any "gas" in general.

When 3 ounces of *magnesia* were heated to very high temperatures – a process called calcining – the solid became red hot. Allowed to cool, the solid residue lost more than half its weight, ultimately measuring "an ounce, three drams, and thirty grains." A dram is one-eighth of an ounce and there are 60 grains to a dram. Only a small portion of the volatiles driven off was water, collected and weighed as 5 drams. Black observed that

> Chemists have often observed, in their distillations, that part of a body has vanished from their senses, notwithstanding the utmost care to retain it: and they have always found, upon further inquiry, that subtile part to be *air*, which having been imprisoned in the body, under a solid form, was set free, and rendered fluid and elastic by the fire.

In modern terms, 7.8 g of magnesium carbonate ($MgCO_3$, *magnesia)* upon heating to high temperature yielded 3.4 g of product magnesium oxide (MgO) in comparison to an anticipated 3.7 g if (a) both starting material and product were pure, if (b) the decomposition reaction were complete, and if (c) the measurements were accurate and precise. In a separate experiment, Black dissolved a weighed amount of calcined (decomposed by high heat) magnesia in spirit of vitriol (sulfuric acid). This would produce soluble magnesium sulfate ($MgSO_4$). He then added large amounts of a mild alkali, either sodium carbonate (Na_2CO_3) or potassium carbonate (K_2CO_3), which re-separated, by precipitation, the solid *magnesia* ($MgCO_3$). The regenerated *magnesia* weighed nearly 100% of the original amount, showing almost complete recovery of the original material. The recovered *magnesia* again effervesced with acids, bubbling off "gas" just as readily as did the original material. Black concluded that the "air" given off in acids is "fixed" in the *magnesia*, released upon strong heating, and returned by combining with something in the alkali, the solid carbonates as we now know them. Hales had previously shown that gas could be released from the mild alkalis upon the addition of acid. Black's careful quantitative measurements demonstrated that the "air" or "gas" is fixed in the solid *magnesia* and lost upon heating, justifying the label "fixed air" for the fluid substance, now known as carbon dioxide. Black's experiments showed, for the first time, that a gas could combine with a solid, something thought impossible until then. Among the many other experiments on "fixed air" performed by Black was a very simple one in which he demonstrated that a candle would not burn in it, dramatizing the idea of his label "fixed air."

Joseph Black

As an aside, the Oxford English Dictionary has some further interesting citations from the early to mid-1700s of references to *choke-damp* and *styth* (or *stythe*) as present in coal mines and elsewhere. The terms were used for CO_2-laden gas and continued in use up to the close of the nineteenth century.

CAVENDISH

In 1766, Henry Cavendish, a British scientist born in Nice, France, published his first article with the Royal Society on "Factitious Airs," detailing a series of experiments that referred to artificially produced substances. Among many of his discoveries, the one for which he is most known is the discovery of hydrogen gas. In later years, he also was the first to determine that oxygen was 20% of air. But for us reflecting briefly on his science now, his experiments determined the density of carbon dioxide gas. An ancillary contribution was his noteworthy observation that the gas could extinguish flames.

HAMILTON

Sir William Hamilton, an architect and British envoy to the court of Naples, became an expert in the actions of earthquakes and volcanoes while living in Italy. In 1770, he wrote to the Royal Society and had published in the proceedings some "Remarks upon the Nature of the Soil of Naples, and Its Neighborhood." In referring, at first, to records of the eruption of Vesuvius in 1631, Hamilton writes

> The nature of the noxious vapours, called here *mofete*, that are usually let in motion by an eruption of the volcano, and are than manifest in the wells and subterraneous parts of its neighbourhood, seem likewise to be little understood. From some experiments very lately made, by the ingenious Dr. Nuth, on the *mofete* of the Grotto del Cane, it appears that all its known qualities and effects correspond with those attributed to fixed air. Just before the eruption of 1767, a vapour of this kind broke into the king's chapel at Portici, by which a servant, opening the door of it was struck down. About the same time, as his Sicilian majesty was shooting in a paddock near the palace, a dog dropped down, as was supposed, in a fit; a boy going to take him up dropped likewise; a person present, suspecting the accident to have proceeded from a *mofete*, immediately dragged them both from the spot where they lay, in doing which, he was himself sensible to the vapour; the boy and the dog soon recovered.

Hamilton smartly recognizes the intimate relationship between Joseph Black's fixed air, first described 16 years earlier – our carbon dioxide – and volcanic eruptions.

PRIESTLEY

Joseph Priestley was born near Leeds, England, in 1733, trained for the ministry, and became ordained before his 30th birthday. He had no formal training in science but was self-taught. On a trip to London, he met Benjamin Franklin and was enticed into a deep interest in electricity. Priestley published his first scientific paper "The History of Electricity" in 1767. Very soon, he launched a prolific career studying gases, studies that actually began with experiments on carbon dioxide. This departure from his interest in electricity was due to his observations of the Meadow Lane Brewery neighboring his living quarters in Leeds. Priestley recognized that the brewery offered an abundant supply of fixed air and began his studies by trying to accumulate it using the gas-collecting device originally designed by Stephen Hales. The problem, though, was that fixed air was rather soluble in the water over which it was supposed to be collected. Priestley's first publication in chemistry was on how to carbonate water, mimicking sparkling mineral waters and restorative spas. For this simple precursor to the carbonated beverage industry, the Royal Society awarded him the prestigious Copley Medal. The title page of his publication is reproduced here.

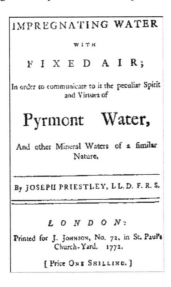

IMPREGNATING WATER

WITH

F I X E D   A I R;

In order to communicate to it the peculiar Spirit
and Virtues of

Pyrmont   Water,

And other Mineral Waters of a similar
Nature,

By JOSEPH PRIESTLEY, LL.D. F. R. S.

L O N D O N:
Printed for J. JOHNSON, No. 72, in St. Paul's
Church-Yard. 1772.

[Price ONE SHILLING.]

Priestley, despite his lack of formal scientific education, was exceedingly clever. To circumvent the loss of soluble carbon dioxide while being collected over water, he devised a variant apparatus using liquid mercury in place of water. Liquid mercury does not dissolve carbon dioxide.

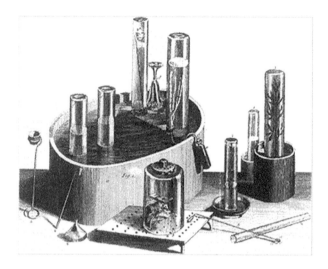

In 1774, Priestley heated a sample of *red calx* and "discovered an air five or six times as good as common air." Red calx is mercuric oxide, HgO, one atom of mercury (element 80) combined with one atom of oxygen. Its decomposition allowed Priestley to collect the nearly pure oxygen gas which proved to be insoluble in water, enhanced the burning of a candle, and supported a mouse sealed within an airtight chamber as shown in the illustration. It is for the discovery of oxygen that Priestley is most widely known, although he didn't recognize its role in chemistry at the time, a gap that his contemporary, the Frenchman Antoine Lavoisier, filled.

Joseph Priestley

We can proceed with carbon dioxide now that its discovery has been accommodated.

4 Structure

Everything has its beauty but not everyone sees it.

CONFUCIUS

CO$_2$

Carbon dioxide is a compound, a molecule containing more than one kind of atom. It consists of one carbon atom and two oxygen atoms. There are other compounds of carbon and oxygen that are known, some stable, some very reactive. The second most common carbon–oxygen compound is the infamous toxic gas, carbon monoxide, with one carbon and one oxygen.

Some of what we know and need to know about carbon dioxide is its physical science, mostly its structure. Carbon dioxide is electrostatically neutral; that is, it is uncharged. But this does not mean that there are no regions of the molecule that possess charge, only that those positive and negative charges add up to zero. There is a property of atoms, when combined in molecules, that indicates the atom's relative tendency to attract valence electrons to it and thereby acquire a slight negative charge, leaving a deficit of charge or buildup of positive charge. It's called the atom's electronegativity. The more electronegative an atom, the more the atom will pull electrons toward it, albeit only slightly. Oxygen is one of the most electronegative of all elements.

There are five conceivable bonding arrangements of one carbon and two oxygens in a single molecule. Representing the chemical bond by a connecting bar and the atoms by their appropriate alphabetical symbols, the possibilities, all planar, are shown here.

O-C-O   either linear or bent

O-O-C   either linear or bent

   triangular

One of the properties of a molecule determined by its geometry and the electronegativities of its constituent atoms is called its polarity. Actually, it's called the electric polarity. In a sense, it is similar to the magnetic polarity of our planet Earth with its north magnetic pole and south magnetic pole. If a molecule has a north electric pole (with buildup of negative charge) separated from a south electric pole, the molecule would possess two poles, a *dipole,* and be said to be a polar molecule. Of course, the molecule doesn't know north from south. All that's needed is a separation of negative charge at one region from positive charge in another region. The greater the amount of charge and the greater their separation, the greater is the polarity of the molecule. Carbon dioxide is known from its behavior to be a non-polar molecule. It does not have a dipole. This fact alone eliminates all the possible structures for carbon dioxide except the linear O–C–O. The linear structure, because of its symmetric form, has the center of negative charge and the center of positive charge at the same location. Each of the other possible structures would be polar. There is other evidence that the linear, symmetric geometry is correct as well, and there is no contradictory evidence.

Calculations, beyond our focus here, can indicate where the electrons are in the carbon dioxide molecule. Shown here in a "ball and stick display" are the partial electrostatic charges (expressed in units of the charge of a single electron) at each atom calculated from a theoretical description.

-0.54   +1.08   -0.54

Although each carbon–oxygen connection itself is polar in the sense that there is a positive pole and a negative pole, the two carbon–oxygen dipolar arrangements oppose each other. The effect cancels out, giving carbon dioxide a non-polar character. There is no center of negative charge separated from a center of positive charge. The center of negative charge is exactly midway between the oxygen atoms which is where the center of positive charge falls. Finally, using the same theoretical approach that yielded the electrostatic charges on carbon dioxide, we can construct an outline that envelops most of the valence electron distribution in the molecule. Red on both ends indicates a concentration of negative charge. The colors correspond to a gradation in charge swinging to the center blue for positive charge.

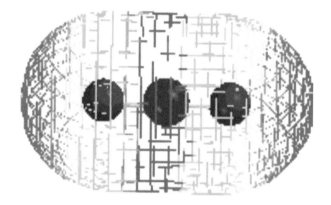

The importance of the charge locations and magnitudes on the carbon dioxide cannot be over-emphasized. Essentially, all physical properties of the molecule depend in subtle ways on the charges. Since like charges repel and opposite charges attract, behavior which involves the interaction of one carbon dioxide with another carbon dioxide or with any other structure involves the relative orientations of the molecules at contact and whether or not they involve attractions or repulsions. Boiling points, pressure of the vapors, solubility, and other behaviors all derive from the picture.

Conventionally, a molecule such as carbon dioxide is pictured as if the atoms were rigidly locked into place by something pictured to represent the chemical bond. While there certainly is a quintessential aspect of molecular structure called the chemical bond, it is in no way supposed to be constructed as a rigid arrangement. The bond is somewhat slack and allows the atoms to jostle around slightly, a motion referred to as vibration. Depending on how many vibrations per second, how fast they are, and on the masses of the constituent atoms, one can imagine there will be a vibrational energy associated with the molecule. You would think that if the temperature were reduced, the molecule could give up vibrational energy. This is certainly so. But unlike the everyday concept that if we cool the molecule enough, the motion should stop completely, well-established theory tells us that such motion cannot stop. There will always be a residual vibrational motion even at the absolute zero of the temperature scale, −479°F or −273°C. The energy associated with the residual motion is called the *zero-point energy* and is a direct consequence of an uncertainty principle (or indeterminacy principle) recognized first by the German theoretical physicist Werner Heisenberg during the mid-1920s.

Vibrational motions in carbon dioxide can only increase in nearly equal jumps of energy, like bounding up a staircase, each step representing a higher and higher vibration frequency. The energy can be provided by light – electromagnetic radiation – that occurs in the infrared region of the spectrum. Infrared light has slightly less energy than that linked with visible light (and much less than that of ultraviolet [UV] light).

If we bathe carbon dioxide with infrared radiation of different frequencies, the maximum effect occurs at a wavelength of ~15 micrometers, corresponding to a frequency (of light's oscillating field) of about 20 trillion cycles per second (20,000 gigahertz). A microwave cooking range operates at about 2.5 billion cycles per second.

There is an additional key requirement here. For the oscillating electric field of light to be effective in producing vibrational jumps there must be an oscillating dipole in the absorbing molecule. This is a requirement for the process we're discussing, although we won't attempt to explain its origin. But carbon dioxide has no permanent dipole as we pointed out already. However, Heisenberg's uncertainty principle guarantees that carbon dioxide must vibrate with at least its zero-point vibrational energy in each of its several modes of motion. It is therefore guaranteed to have an oscillating dipole in bending and asymmetrical stretching vibrations. Even though the average dipole is zero, there is a flip-flopping dipole.

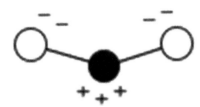

The absorption of light corresponds to an energy match that would allow a jump in the vibrations from 10 to 30 trillion cycles per second, a wavelength of 15 micrometers corresponding to the energy* needed to accomplish this. Because the remaining vibrational mode, the symmetrical stretch, does not involve an oscillating dipole, the molecule will *not* absorb the infrared light that would otherwise perfectly match the energy needed to boost the vibrational frequency for this mode. Incidentally, this is the precise reason that molecular nitrogen (N_2) and molecular oxygen (O_2) do not absorb infrared radiation. Their vibrations, by the very symmetry of the two molecules, do not have oscillating dipoles associated with them. They are transparent to infrared light.

As esoteric as this recent discussion may seem, it is actually the embedded Heisenberg uncertainty principle that explains why carbon dioxide is so intimately associated with being a greenhouse gas, effectively absorbing infrared radiation emitted outward from Earth's surface rather than it passing directly into space. More on that later.

OTHER PROPERTIES OF CARBON DIOXIDE

Carbon dioxide is one-and-a-half times as heavy as air. Since CO_2 doesn't support combustion and is easy to produce, it is used in many fire extinguishers as a way of suffocating a flame, starving the flame of oxygen.

The gas may be liquefied by cooling to −18°C at a high pressure of about 20 atmospheres or, alternatively, at "room temperature" 21°C and a still higher pressure of about 67 atmospheres.

At temperatures greater than 31°C and pressures greater than 74 atmospheres, carbon dioxide will have been compressed to the point where liquid and gas are indistinguishable and one has

---

* 15 micrometers wavelength is equivalent to 1.3×10^{-20} joules or 3,200 billionth of a billionth calories for one molecule or 44 calories per gram.

what's just called a *fluid*. The conditions under which this first occurs are referred to as a critical point and so the fluid phase is called *supercritical* carbon dioxide. Commercial uses of supercritical carbon dioxide are growing extremely rapidly. The fluid is recyclable and easily purified. It has replaced ultrapure water in computer chip manufacturing and is used to clean electronic parts and mechanical parts. Chlorinated organic solvents such as trichloroethylene (TCE, also known as trichlor) used in dry cleaning are toxic, but are being replaced with supercritical CO_2. The fluid is also used as a solvent in a variety of specialty situations, the most well-known of which is the removal of caffeine from coffee.

OTHER CARBON–OXYGEN MOLECULES

Among other carbon–oxygen compounds, there is, of course, carbon monoxide (CO). Less well known are small clusters of these molecules. Combinations of two and three carbon monoxides and carbon dioxides form molecules whose formulas could be written as C_2O_2, C_3O_3, C_2O_4, and C_3O_6. A carbon suboxide has been known for years and has the formula C_3O_2. Measurements of light passing through interstellar molecular clouds have been used to identify dicarbon oxide (ketene) and tricarbon oxide, C_2O and C_3O, respectively. Under normal conditions, in the atmosphere or everyday environment, these would react and disappear.

Finally, the tetra-hydroxide of carbon exists *in theory*. It would be the product of reacting carbon dioxide with water:

$$CO_2 + 2H_2O \rightarrow C(OH)_4$$

And if favored as a reaction, the tetra-hydroxide would completely change the nature of the world. Carbon dioxide and water are discussed in Chapter 10. Further reaction of this hypothetical substance with basic elements leads to the formation of *orthocarbonates*, a known example of which would be sodium orthocarbonate, Na_4CO_4.

PHASES

Elemental carbon has two very well-known phases, assemblies of uniform properties, distinct and separable from others: graphite, the writing substance in a so-called lead pencil, and diamond, which is formed in certain geological environments deep beneath Earth's surface under very high pressure. A third form has achieved notoriety recently and is associated with the 1996 Nobel Prize in Chemistry. The form is *buckminsterfullerene*, or "bucky ball."

Carbon dioxide is in the gas phase…under ordinary circumstances. However, if you drop the temperature to −78.5°C (−109.3°F), carbon dioxide freezes into a solid phase. That solid phase is commonly referred to as *dry ice*. Warm the dry ice, and it converts back into the gas. Unlike most other solids, it does not melt into a liquid first. (The conversion of a solid into a vapor without melting first into a liquid is referred to as *sublimation*.) The liquid phase of carbon dioxide does exist, but only at pressures above five atmospheres. And at that pressure, the temperature must be no lower than −56.6°C.

At even higher pressures, solid carbon dioxide can take on different forms. One form is known as *carbonia* and was discovered recently, in 2006. It is analogous to the glass form of silicon dioxide. Multiplicity of solid structures is a fairly common phenomenon. Ice, to date, has nine different solid forms (one of which was the basis of Kurt Vonnegut's novel *Cat's Cradle*).

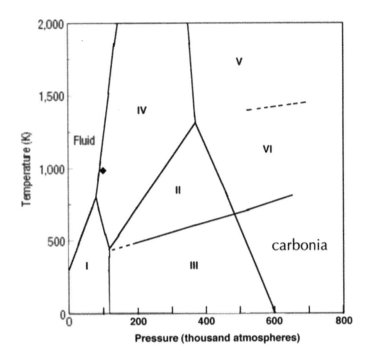

In this diagram of temperature *versus* pressure with the various solid phases indicated by Roman numerals, the solid lines indicate where two phases would co-exist, like ice and water at 0°C. Note that if carbon dioxide were just above 100,000 atmospheres and 1,000 K temperature (the diamond on the figure), it would be CO_2 (solid) phase IV. If the pressure were reduced, moving horizontally to the left on the diagram, the solid would become liquid CO_2. That is, it would melt without the application of heat. Most material, water ice being an exception, behaves like carbon dioxide in this way. But since water ice is the substance we have the most experience with, the aforementioned melting phenomenon seems counter-intuitive to our experience.

Carbon dioxide may take several other distinct solid forms. Studies* on these have recently advanced our knowledge but are still not completely resolved. Among observations are two that are particularly interesting. At high pressure, carbon dioxide may convert to a "superhard" solid structure. Additionally, there is the possibility that solid carbon dioxide may exist in Earth's mantle as a structure similar to its analog, quartz: SiO_2. The more CO_2 is studied, the more surprises are revealed.

---

* J. Sun et al., "High-pressure polymeric phases of carbon dioxide," *Proc. Natl. Acad. Sci.* 106, 6077 (2009).

5 Radiocarbon and Its Dioxide

Carbon-14, the long-lived carbon isotope, is the most important single tool made available by tracer methodology, because carbon occupies the central position in the chemistry of biological systems.

MARTIN KAMEN
Co-discoverer of carbon-14

Longest lived among carbon's radioactive isotopes is carbon-14. This is a naturally occurring radioactive substance and, over the years since its discovery by Americans Martin Kamen and Sam Ruben in 1940, has served as a powerful analysis tool in a variety of applications. Those applications are part of the carbon dioxide story and we will explore some of these after we look at why there is any carbon-14 around when its average lifetime is about 8,000 years yet our planet is over 4 billion years old.

COSMIC RAYS

The process of removing one or more electrons from an atom or molecule is called ionization. (Adding an electron to form a negatively charged ion is not referred to in this manner.) High energy particles emitted by the sun reach Earth's atmosphere and there cause ionization of atoms and molecules. Many more particles hitting Earth come from outside the solar system, that is, are galactic or extragalactic. They are mostly protons – 90% – with another 9% being alpha particles, and the rest are yet heavier nuclei. (Their energies are so great that their detailed behavior needs to be understood using Einstein's theories of relativity.) In the atmosphere, via nuclear reactions, they can produce showers of other nuclear particles including neutrons. Cosmic rays are ultimately responsible for a variety of phenomena among which are the auroras seen in the more polar latitudes of both northern and southern hemispheres. They influence not only the chemistry of the atmosphere, transformations from one molecular arrangement to another, but some nuclear transformations as well. As far as our book's topic is concerned, cosmic rays provide a fortuitous by-product with many benefits.

PRODUCTION OF RADIOCARBON

The most common atom in the atmosphere is nitrogen, element number 7. Nitrogen exists mostly as a gas whose constituents are two nitrogen atoms chemically bound together: molecular nitrogen, N_2. All but about 0.4% of naturally occurring nitrogen atoms have mass number 14. The mass number implies that in addition to the seven protons in the nucleus, there are also seven neutrons for a total of 14 elementary particles. A rare collision in the atmosphere of a fast neutron from cosmic rays with the nitrogen-14 nucleus sometimes happens in such a way that a proton is knocked out of the nucleus and the neutron remains. You could picture this akin to what happens when a fast billiard ball strikes head-on one that is stationary. The struck one recoils quickly in the original direction and the incoming ball remains behind in its stead. This simplified picture means that the struck nucleus with its components of seven protons and seven neutrons is converted into a product with six protons and eight neutrons. That final combination is the isotope carbon-14.

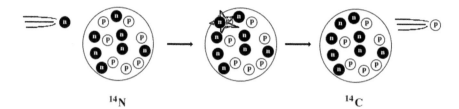

Even more often, a slow neutron is captured to form ^{15}N, depositing enough energy that a proton is subsequently ejected a very short time later to produce the ^{14}C.

Carbon-14, sometimes called just *radiocarbon*, is practically identical in its chemical behavior to carbon-12 and carbon-13 but is unstable with respect to radioactive decay at a very slow rate. A carbon-14 lives an average of over 8,000 years before being transformed by a spontaneous nuclear transmutation. That is by far long enough for the carbon-14 to become chemically incorporated into atmospheric carbon dioxide, for example, and to have its share of the myriad of chemical processes that occur as part of the so-called carbon cycle and to serve as a tracer or proxy of those processes. It also means that the vegetation we consume unavoidably contains radiocarbon (unless the plants are tens of thousands of years old). The average human weighing 70 kg is about 12.6 kg carbon. The trace amount of radiocarbon there would give an average of 3100 decays per second, non-stop, and constant. In radioactivity units, we are talking about 85 nanocuries (or 3100 becquerels).

Production and decay of carbon-14 are more or less balanced in the atmosphere at the present. The average production rate is 2.2 atoms per second for every square centimeter, 90% from solar cosmic rays, the remainder mainly galactic. An equilibrium amount of 15 decays per minute for every gram of carbon in the atmosphere is obtained. The atmosphere contains a steady total amount of about 6 megacuries of radiocarbon – naturally.

ISOTOPE EFFECTS

In the "Production of Radiocarbon" section, there was a subtle hedge in the comment on the chemical behavior of the different carbon isotopes, saying they were "practically identical." There are, in fact, very slight but definitely measurable and understandable differences in the chemistry of different isotopes of the same element due to their slightly different masses. A more thorough explanation of the origin of this *fractionation* effect is deferred to Chapter 20, but presenting some observations on the effects of different isotope masses make sense now. In working with carbon-12 (the 99% abundant, stable isotope) and switching to carbon-13 (also stable) properties will alter very slightly. The $^{13}C/^{12}C$ abundance comparison is expressed relative to a standard sample in terms of percent difference of that ratio from that reference standard ratio. (Professional researchers who study this topic, instead of *percent* deviation, use *per mil*: ‰ = per mil or parts per thousand.) Atmospheric carbon dioxide has a fractionation of −0.9%. If the standard had exactly 1% ^{13}C (ignoring the fact that it is not precisely 1%), a −0.9% fractionation means that the air sample instead has* 0.991% ^{13}C. The "−0.9%" deviation indicates that there is relatively less ^{13}C than in the standard. Carbon in desert cactus, to pick an example, has a fractionation of +1.7%. Leaves from trees have typical fractionations of +2.7%. The three examples quoted actually have a range of measured values. There are also many more examples of representative fractionations for carbon (and oxygen). Not unexpectedly, ^{14}C will have an isotope fractionation effect that is even greater than that for ^{13}C (since its mass differs from that of ^{12}C by more than does that of ^{13}C).

---

* The standard ratio is 1%/99%. The sample ratio gives a fractionation difference $(x − 1/99)/(1/99) = −0.9$ in % or $x = 0.01001$ the ^{13}C-to-^{12}C ratio in the sample. Since those mass 12 and 13 isotopes add to 100%, the ^{13}C becomes 0.991% of carbon.

RADIOACTIVE DECAY OF CARBON-14

Carbon-14 is unstable. It is heavier – more massive – than nitrogen-14. Each of these isotopes has a *mass number* of 14, representing the total number of protons and neutrons in the nucleus. But the atoms' actual masses differ by about one-thousandth of a percent. Carbon-14 can change to stable lower mass* nitrogen-14 because a pathway is open to allow that change. The pathway is radioactive *beta-decay* in which the six protons and eight neutrons in the nucleus spontaneously change into seven protons and seven neutrons. Effectively, one neutron has converted into a proton, emitting a negatively charged electron from the nucleus. That electron gets a special name, a beta particle.† Beta particles can be detected with sufficient sensitivity that corresponds to 10,000 year-aged samples. (Rather than exploiting radiation measurements, ^{14}C atoms themselves can be measured by accelerator mass spectrometry, improving the sensitivity enough to samples of age 50,000 years.) The beta particle is identical in all respects except history to an electron. For radiocarbon, the probability of this beta decay, although guaranteed, is fairly low. The rate is a property of the nucleus and corresponds to about a 1.2% chance of decay in a century for any carbon-14 atom.

It is instructive here to follow what happens if we start with a large number of such nuclei. Suppose we initially have 100% ^{14}C. After one century, 1.2% of these will have been lost, becoming product ^{14}N and leaving behind the remaining 98.8% ^{14}C yet unchanged. Over the course of the next century, 1.2% of the latter decay into ^{14}N, leaving behind 97.6% of the original ^{14}C. That is, if 1.2% decay, 98.8% survive the century. We started with 98.8% of the original and if 98.8% of these survive into the second century of elapsed time, that's where the 97.6% figure comes from. The next century would leave 96.4% of the original; the next would leave 95.3% and so on. After one millennium, 88.6% of the original ^{14}C would remain; after two millennia 78.4% would be left. If we graph this behavior, the figure shown results.

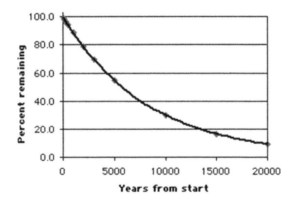

This type of behavior follows what is called an *exponential* curve. What you might notice is that the amount of time it takes for half of the ^{14}C to decay is between 5,000 and 6,000 years. The actual figure is 5,700 years and this time is referred to as the *half-life* of ^{14}C. No matter when you begin a measurement, in 5,700 years half of the ^{14}C will have turned into ^{14}N. Carbon-14's average life is around 8,200 years, a value that is unaffected by anything in the surroundings. After ten lifetimes (82,000 years), 99.98% of the original amount has decayed.‡ A "lifetime" does not imply that the substance is extinguished as if it were living. Rather, the meaning is that this is the average time a nucleus remains unchanged. Some decay earlier, some later…randomly.

---

* Lower mass = lower energy, vis-à-vis $E = mc^2$.
† An additional particle, an antineutrino, is also emitted but that detail is not germane to our discussion.
‡ A legitimate question is the impact of the nitrogen-14 produced. There is so much more nitrogen in the atmosphere that the contribution from radiocarbon decay has negligible effect. Six megacuries of carbon-14 produce (6×10^6) (3.7×10^{10}) nitrogens per second or fewer than 2 g per year. Even millions of years would have minimal effect.

Just to try to illustrate that the half-life of an unstable isotope is a characteristic of that particular isotope, a few other half-lives are worth mentioning. Uranium-238 has a 4.5-billion-year half-life. Naturally occurring (even in "substitute salt," potassium chloride, KCl) potassium-40 has a 1.3-billion-year half-life. Radon-222, the naturally occurring isotope of the radon of health concern in basement gas seepage, has a 3.8 day half-life. Fluorine-18, used in PET scans, has a 110 minute half-life. Recently discovered and named superheavy element (number 118), *oganesson* has a ^{294}Og half-life of 0.0007 seconds.

Because the isotope disappears at a certain rate, the ^{14}C at first accumulates. But then an equilibrium is reached in which the rate of accumulation is balanced by the rate of disappearance. The amount of ^{14}C in the atmosphere would then remain constant. The balance between appearance and disappearance is called a "steady state" situation.

Carbon-14, because of its *nearly* perfect imitation of natural ^{12}C chemistry, can be used to determine age, a fact that was first recognized by American chemist Willard Libby. The technique had proven so valuable to archaeology and paleontology, the study of former life forms, usually *via* fossils, that Libby was awarded the 1960 Nobel Prize in Chemistry for his discovery.

Let us look at a tree as an illustration of the radiocarbon dating method. While the tree is living, carbon dioxide from the atmosphere is used to produce wood, basically the substances cellulose and lignin, containing lots of carbon. Not only does that carbon consist mostly of ^{12}C with about 1% ^{13}C, but also of some tiny amount of ^{14}C that was present in the air, the ^{14}C having been produced by cosmic rays. To a good first approximation, the amount of ^{14}C present in the air has been constant over many millennia. If so, then the ratio of ^{14}C to ^{12}C will be nearly constant in living parts of the tree, older parts having very slightly less ^{14}C than younger parts of the same plant.* In wood where carbon is no longer being incorporated, the ^{14}C/^{12}C *ratio* begins to drop because of radioactive decay. When the tree dies, the ^{14}C/^{12}C ratio drops everywhere, always at the rate determined by the ^{14}C half-life. A measurement then of the remaining ^{14}C through its decay rate directly tells you how much ^{14}C you have in a sample. If you isolate 1 g of pure carbon from newly formed wood, you obtain (approximately) the equilibrium 15 decays per minute. If you have another piece of wood and you do the same experiment, but the decay rate of ^{14}C is measured to be 7.5 decays per minute[†] in a gram of pure carbon, this is an indication that the ^{14}C has not been replenished for an amount of time equal to one half-life: half has decayed and half remains as seen in the graph. The sample is 5,700 years old. That is how long it has been since growth ceased. The age of that piece of wood is 5,700 years.

AGE CALIBRATIONS

That simple illustration of radiocarbon dating began with the proviso that atmospheric radiocarbon abundance is and was always the same. Cosmic ray intensity bringing about radiocarbon production has not been absolutely constant though. And we're not talking particularly about the fluctuations associated with the 11-year sunspot cycle which amounts to about 0.1%. More significant fluctuations, including alterations in Earth's magnetic field, which can deflect cosmic rays, force corrections onto the interpretation of the radiocarbon-determined ages. Indeed, it was the Dutch physicist Hessel de Vries who in 1958 was the first to demonstrate that radiocarbon ages were imperfect.

Fortunately, for the most recent seven or so millennia, such corrections are confidently determined through a calibration using "dendrochronology," a technique invented by the American astronomer Andrew E. Douglass just before the start of the twentieth century. Douglass discovered the apparent correlation between climate and plant growth from evidence in tree rings. Experience and common sense tell us that thick rings correspond to productive growing seasons. History

---

* From earlier, a century difference in age corresponds to a 1.2% difference in ^{14}C content.
† There cannot be a fractional decay, but 7.5 decays per minute means the same as 15 decays in 2 minutes, for example.

of interest in tree rings* extends back as far as Aristotle's student and successor, Theophrastus (370–285 B.C.E.), considered by some to be the "father of botany." Leonardo da Vinci recognized the relationship between tree rings, moisture availability, and past weather. In the mid-1700s, Carolus Linnaeus ("father of modern taxonomy") noted that tree rings record historical weather patterns and called rings "chronicles of winters." Charles Babbage ("father of the computer") in 1837 suggested that overlapping patterns of tree rings could enable construction of ancient chronologies.

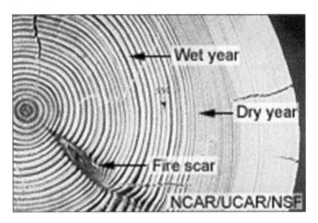

Andrew Douglass demonstrated that there were matching "records" in tree rings from collections of trees. He used overlapping growth patterns as a dating technique. A schematic representation of the technique is diagrammed here.

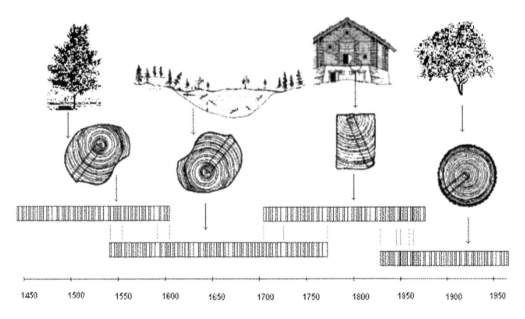

Over the years, this idea has been extended to include radiocarbon data from layered lake sediments called *varves*. Their similarity to tree rings was noted by Dutch–Swedish geologist Gerard de Geer in the late nineteenth century. Establishing the chronology of the layers, though, is complicated by the occasional double layers in one year or missing layers due to the vagaries of seasonal growth.

---

* R. Wimmer, "Arthur Freiherr von Seckendorff-Gudent and the early history of tree-ring crossdating," *Dendrochronologia* 19, 153 (2001).

In a similar vein, studies of coral, which also have growth rings (illustrated here), have proven useful, as have radiocarbon measurements of stalagmites in limestone caves. In the latter case, radiocarbon readings are shifted from the ages of the stalagmite deposits because of the time expended to produce the required carbonate from overlying organic-rich soil. Furthermore, the porous nature of the stalagmites allows "young" carbonate to seep into "old" accumulations and blur the age information.

The various data points in the figure show the calibrated "radiocarbon" age along the vertical axis, most often slightly underestimating the true age. There are slightly different calibrations for the two hemispheres. Time resolution is five years for just over the first 10,000 years. Radiocarbon age is what is extracted from the portion of carbon that is ^{14}C remaining in the sample, calculating the time needed to drop from the current value to that measured one. The horizontal axis is the actual age of a sample – determined by other, independent measurements or records – whose radiocarbon content will be measured. For the first 10,000 years, the age comes from dendrochronology. Subsequently, ages come from other radioactive isotope dating techniques that do not depend on constant cosmic ray intensities over time and also from measuring sediment layers in lakes and growth rings in corals. The straight line shows where measurements of carbon-14 would be expected to directly yield accurate ages if no corrections proved necessary. Prior to 10,000 years ago, the general interpretation is that cosmic ray intensity was less than more current levels, producing less ^{14}C and giving younger ages. A near shutdown of Earth's magnetic field several tens of thousands of years ago is hypothesized as an explanation. The proposal is supported by magnetic measurements in deep sea sediments and by ^{10}Be measurements (another radioactive species generated by cosmic rays). Naturally, the amount of nitrogen in the atmosphere from which the radiocarbon is generated is understood not to have fluctuated over these millennia.

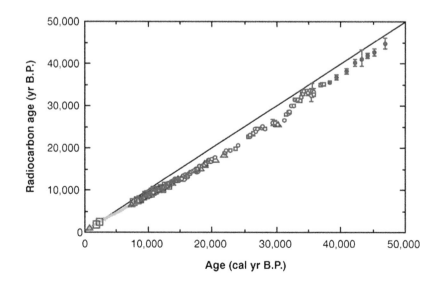

DEVIATES

Fluctuations in cosmic ray intensity, and thereby the production of radiocarbon, can occasionally have confounding effects. For example, in the calendar age *versus* radiocarbon age calibration graph, the ^{14}C "clock" is seen to have stalled for several thousand years around 26,000–31,000 years ago (the curve flattens) possibly and probably due to below average radiocarbon production. Prior to that, starting around 31,000 years ago, the "clock" accelerated, presumably reflecting above average production of radiocarbon.

Standards are needed for precise and accurate (true) age determinations. By combining the ^{14}C measurements and sequencing of tree rings through successive seasons in trees whose lives over-lapped at some identifiable point or points as shown in the earlier schematic diagram, calibration can be accomplished. The technique enables determination of a calibration extended over many millennia. The age of a tree wood ring is determined by counting rings backward from the present (or from a known date), then matched with the age determined just from ^{14}C content of the ring to establish a calibration or correction. An example of calibration that covers just the last several hundred years in very precise detail is shown here.

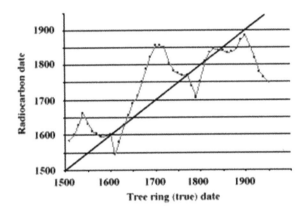

On the horizontal axis is the true date of the sample being analyzed for radiocarbon in tree rings determined by *dendrochronology* (counting and assigning rings). On the vertical axis is the age calculated from measured ^{14}C content and the ^{14}C-to-total-carbon decay factor. The straight line through the intersection of the two axes indicates where the data would have been if there were no corrections necessary, that is, if the radiocarbon date exactly mirrored the true date. You see that corrections are necessary. For example, the tree ring that grew during the year 1600, when analyzed for radiocarbon to determine its age, gave (on the horizontal) 1600 also. But for the tree ring that actually grew in 1700 according to the tree ring counting result on the horizontal axis, an apparent age from radiocarbon dating is 1850 (for the vertical axis) implying that too much radiocarbon is present (150 years' worth) compared to the correct age. For tree rings that date at 1850, this means there is 1.8% too much carbon-14 in the uncorrected radiocarbon description.*

OTHER COMPLICATIONS

If an undetermined sample of wood, say, has its $^{14}C/C$ level measured and yields a radiocarbon age of 150 years (prior to the reference year, 1950) or a date of 1800 by extrapolating backward using the known half-life, the calibration data show that the true age in this case is ambiguous. The measured $^{14}C/C$ level on the calibration would agree with dates of 1675, 1725, 1820, and 1925 equally well. These correspond to ages both older and younger than the radiocarbon level indicates.

---

* Recall from earlier that a century corresponds to a 1.2% carbon-14 change.

What is displayed next is the standard *correction* for radiocarbon in tree rings determined by dendrochronology and standard tree samples from the previous illustration.* The peak at 1.8% for the correction factor corresponds to the year 1700 as just discussed. Note, too, that at 1600 the correction happens to be 0%.

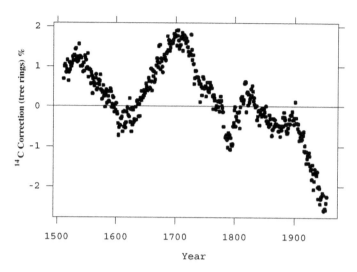

An even more detailed study of deviations in atmospheric radiocarbon production was initiated by a 2012 study of tree rings in cedar by a team of scientists in Nagoya, Japan.[†] The scientists found a sudden jump in radiocarbon associated with the year 775 CE using tree ring dendrochronology with one to two year time resolution, a considerable improvement in what is used in the official calibration standards that tie radiocarbon ages to actual ages. Questions about the cedar wood origin and date naturally arose and answering those questions proved to provide additional valuable insight. Suggestions about the cause included a strong solar burst of cosmic rays; a supernova gamma ray burst; the impact of a comet on the sun producing a coronal shock; and a comet whose high nitrogen content would have been a source of radiocarbon via nuclear reactions. The following year, identical dating and amplitude effects were found in German oaks thousands of miles from Japan in studies by two independent laboratories.[‡] Also identified in the samples were jumps in ^{10}Be and ^{36}Cl radioisotope abundances, proxies for cosmic rays. The timing of the phenomenon matches a cluster of *aurora borealis* reports in Chinese chronicles. Calculations implied that, if a comet impact were responsible, the object would have to have been greater than a kilometer in size and the effect would have been basically confined to one hemisphere of the planet. A 2014 international collaboration[§] performed yet another series of measurements, this time on larch trees from northern Siberia and bristlecone pine from California. Annual resolution in tree ring counting matches the 1.5% radiocarbon jump accurately and precisely as shown in the figure for the five global investigations.

---

* In actuality, isotopic fractionation of ^{14}C is also included in the values displayed. This is just an indication of the very slight differences in how plants distinguish between the two isotopes of carbon being determined, ^{14}C and ^{12}C, namely, there is not a perfect correspondence in the incorporation of CO_2 between both isotopes. The fractionation adjustments are known to be accurate.

[†] F. Miyake et al., "A signature of cosmic ray increase in AD 774–775 from tree rings in Japan," *Nature* 486, 240 (2012).

[‡] G. Usoskin et al., "The AD775 cosmic event revisited: the Sun is to blame," *Astronomy and Astrophysics* 552, L3 (2013).

[§] A. J. T. Jull et al., "Excursions in the ^{14}C record at A.D. 774–775 in tree rings from Russia and America," *Geophys. Res. Lett.* 41, 3004 (2014).

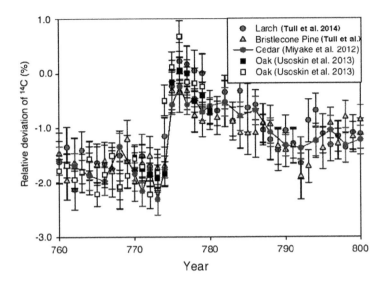

The same international collaboration found identical radiocarbon spikes in Kauri wood from New Zealand and also in corals, dated by counting growth rings, in the South China Sea. Some questions were raised about missing tree rings and erroneous date assignments if growing seasons were slow enough to make a ring hypothetically indistinguishable from neighboring years' rings. But yet another study, this one from pine trees in the Austrian Alps,* seems to obviate such concerns. The samples were from elevations of more than 2 km where low temperatures should have produced frequent slow growth seasons. Nevertheless, the dendrochronology of the observed radiocarbon spike again matched the year 775 results observed globally. All of these investigations of the phenomenon that occurred well over a millennium ago serve to lend strong confidence to radiocarbon chronology studies. The guilty solar event is arguably the strongest such event in over 11,000 years.

Around the year 775, there would have been an equilibrium amount of 815 kg of ^{14}C in the atmosphere with a normal production rate of about 6.7 kg per year. The spike seen implies a *two-fold increase in production* at that time (diluted to a 1.5% effect because of the much higher amount already present).

ANTHROPOCENE

Since about 1900, the $^{14}C/C$ ratio has been significantly disrupted by two man-made interferences. The industrial revolution started in the late 1800s. By 1900, very large quantities of fossil fuels, those that had been buried since antiquity – coal, natural gas, and oil – were burned. Since coal is many millions of years old, any ^{14}C originally present would have decayed completely (as is confirmed by measuring various coal sources). Coal burning would therefore introduce stable C isotopes into the atmosphere beyond normal (natural) sources, diluting the $^{14}C/C$ ratio from its ideally constant value in the air. This effect was recognized by the Austrian chemist Hans Suess and is named after him. Between 1850 and 1950 it is well documented that combustion of carbon-containing materials added 10% excess – radioactively "dead" – carbon dioxide to the atmosphere above and beyond what had been historically occurring. This should have been reflected as a 10% drop in the $^{14}C/^{12}C$ ratio. Yet interestingly, the radiocarbon studies of tree rings illustrated previously for the years 1500–2000 showed that the reduction was only 2 to 3%, evident in the dip of the data near the right edge of the graph. The most rational explanation for the significant discrepancy is that

---

* U. Büntgen et al., "Extraterrestrial confirmation of tree-ring dating," *Nat. Clim. Change* 4, 404 (2014).

there must be a very effective carbon dioxide repository or "sink" sequestering the excess carbon dioxide production, a sink that is larger in its effect than is the atmosphere (ruling out causes by living species). Soil humus is a large reservoir of carbon, but the rate of exchange is too slow to account for the observed change in step with the industrial revolution. The large and very influential carbon dioxide sink is the ocean, as will be explored in subsequent chapters.

The other human intervention in the radiocarbon ratio came about during the period beginning in the 1950s when atmospheric testing of nuclear weapons was, well, prolific. Most nuclear detonations, but not all, were conducted in the northern hemisphere by the United States and by the Soviet Union. These atmospheric tests produced copious bursts of neutrons which would increase the amount of ^{14}C present over and above that due to production by cosmic rays. In the graph we show tree ring determinations* of ^{14}C from Austria, representing the northern hemisphere in which most of the testing was done, from New Zealand, as proxy for the southern hemisphere, and from Ethiopia. Note immediately that the vertical scale, representing departures from unperturbed radiocarbon content expectations, approaches 100%. The method employed collection of carbon dioxide followed by radioactivity assay.

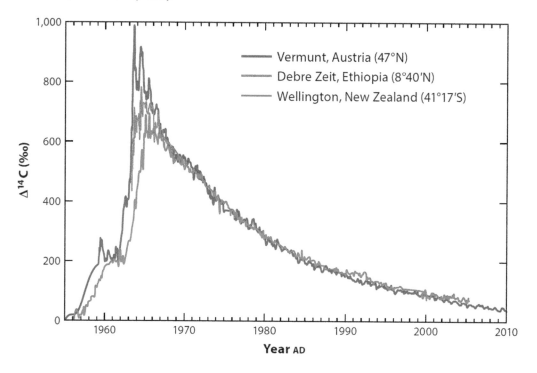

The phenomenon has been taken advantage of as a means of studying stratosphere–troposphere exchange, variable ocean circulation, and sea–air exchange of carbon dioxide. The difference between the two global hemispheres is *not* as vivid as would be expected from the northern hemisphere sites of the bomb-produced ^{14}C. That is an indication of the extent of mixing of air with its CO_2 between the two hemispheres. Testing by various nations began with fission bombs in the 1950s and in the 1960s continued with the introduction of thermonuclear (hydrogen fission–fusion–fission) bombs producing even more marked spikes. The data are zeroed from the 1940s levels as an indication of where the ^{14}C "correction" would track in the absence of these recent man-made perturbations. By looking at these curves, you are able to discern that it took very roughly 15 years

---

* K. Dutta, "Sun, ocean, nuclear bombs, and fossil fuels: Radiocarbon variations and implications for high-resolution dating," *Annu. Rev. Earth Planet. Sci.* 44, 239–75 (2016).

for the ^{14}C peak to fall to half its value. In comparison to the ^{14}C half-life of 5,700 years, the shorter disappearance time is a clear indication of the existence of physical and biological mechanisms for removing atmospheric carbon dioxide. It does *not* indicate that half of the carbon dioxide is removed in 15 years, but that half is turned over, cycled to an extent with fresh carbon dioxide, at that rate.

One of those carbon dioxide sinks – no surprise here, as we mentioned – is the ocean. Radiocarbon measurements of the oceans themselves have been made. Displayed in the accompanying figure is ^{14}C from calcium carbonates* in coral reefs that can similarly be dated by counting age rings in the coral. Coral grows only at relatively shallow marine depths. The radiocarbon spike from these coral near the surface proves to serve as a tracer of carbon dioxide in the oceans and of the oceans themselves. In the figure, previously described corrections derived from tree rings are indicated by the broad gray slightly wavy line up to about 1950. The 5% difference from coral carbonate radiocarbon levels has to do mostly with different isotopic fractionation in coral *versus* that in trees. That is, typical isotope fractionation for coral, distinguishing ^{14}C from ^{12}C, averages –5%. What is shown by the data points is the deviation of radiocarbon content from what is ideally expected for the age of the coral analyzed. In the 1950s and 1960s (in the figure's bracketed region), nuclear testing added ^{14}C to the atmosphere peaking in 1963, but the effect doesn't peak in coral until years later, in the 1980s. Note similarities and differences with the previous graphs from tree ring studies. A bit of thought lets us realize that the delayed peaking is just a reflection of the delay due to the non-instantaneous rate of exchange (recall the 15 year "half-life" for atmospheric weapon-produced ^{14}CO$_2$) of atmospheric radiocarbon with the oceans and the time needed for the slow growth of coral. The negative correction for older coral is also noteworthy. Radiocarbon levels of surface waters in the oceans appear 400 years older than the overlying atmosphere (that feeds trees) due to mixing – dilution – with deeper, older waters and can serve to quantify the rate of such mixing.

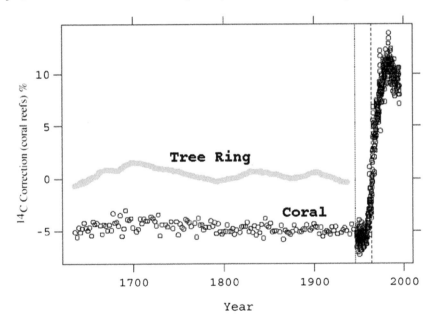

MODERN TWEAKS

Recent atmospheric measurements of the amount of radiocarbon in atmospheric CO$_2$ were slightly lower in the northern hemisphere compared to the southern hemisphere. This is reasonably ascribed

---

* Carbonates, discussed more thoroughly in Chapter 13, are combinations of carbon dioxide with metal oxides.

to differences in non-radioactive carbon dioxide dilution from the burning of fossil fuels in the north *versus* the south. Nevertheless, part of the effect is tempered by the release of extra radiocarbon from nuclear power reactors, prevalent in the northern hemisphere. The observation is a turnaround from preindustrial levels where tree ring measurements implied less radiocarbon in the southern hemisphere. This latter result, arguably normal, arises from the upwelling of deep marine carbon dioxide, depleted in radiocarbon by decay over long residing time, and subsequently released back to the southern hemisphere air to dilute its carbon-14 concentration.

An intriguing observation reported at the 21st International Radiocarbon Conference held in Paris in 2012* was made of air sampling in Indianapolis, United States. A sharp downward spike in local radiocarbon was spotted. The investigators realized this was due to the famous Indianapolis 500 auto race which had just occurred. But it was not emission from race cars that produced the spike, but rather the 400,000 spectators who drove there and back consuming fossil gasoline and diluting the local ^{14}C readings.

In natural science the principles of truth ought to be confirmed by observation.

CAROLUS LINNAEUS

---

* M. Balter, "Neandertal champion defends the reputation of our closest cousins," *Science* 337, 400 (2012).

6 The Air Today

Our most basic common link is that we all inhabit this small planet. We all breathe the same air.

JOHN F. KENNEDY

AIR PRESSURE

You can't really feel it, but the weight of all the air on top of you pushing down from above is what we commonly refer to as air pressure. Typical pressure of the atmosphere at sea level is a bit less than 15 pounds over a square inch or about a kilogram per square centimeter. This value varies with altitude since, obviously, the higher you go, the less air is above weighing down upon you. In the other direction, a whale at depth 1 km under water is subject to a pressure about a hundred times greater than atmospheric pressure.

Since the atmosphere is pushing down with 1 kg of weight on each square centimeter of Earth's surface, we can figure out the total weight or mass of the atmosphere. Earth's diameter is some 8,000 miles (13,000 km). The area of a sphere whose diameter is 8,000 miles is 200 million square miles or 5.1×10^{14} m^2 (or 5.1×10^{18} cm^2). If there is a kilogram of air pushing down on each square centimeter, the total mass of the atmosphere is then 5.1×10^{18} kg. That there are mountains and valleys with corresponding decreasing or increasing air pressure, respectively, complicates the calculation. But ignoring their effect doesn't change the final value enough to concern us.

AIR COMPOSITION

Jean Baptiste Boussingault was a nineteenth century French agricultural and analytical chemist. Born in Paris, he spent time in Colombia, South America, and served with Simon Bolivar during their revolutionary times in the 1820s. While there, he sampled gas emissions from fumaroles near one of the many volcanoes and found carbon dioxide to be present. On returning to France, he made careful measurements of the atmosphere and was the first to determine that carbon dioxide in air was between 280 and 310 parts per million by weight. He did this in the 1830s.

Jean Baptiste Boussingault

Air, we all know, is composed mostly of nitrogen. Oxygen comprises only about one-fifth of air. Argon is about 1%. In terms of how many gas molecules there are in an arbitrary volume, that turns out to be just less than 30 billion billion molecules in a cubic centimeter at sea level at 20°C. The value is both temperature dependent and altitude (pressure) dependent. Water vapor can reach as high as 4%. By the time we reach airline cruising altitudes of 10 km (32,000 ft), the number of molecules in that cubic centimeter has dropped to one-third of its sea level value. At 10 times this altitude, the density has dropped precipitously. There are fewer than a millionth as many gas molecules in the cubic centimeter at 100 km altitude as there are at sea level. Furthermore, the composition has also changed. About 3% (30,000 ppm) of the air at 100 km height now consists of *atomic* oxygen (O) in addition to the molecular oxygen (O_2) that we are familiar with. Carbon dioxide at this altitude is roughly 250 ppm. Atomic oxygen is chemically very reactive. Doubling our altitude again, bringing us to 200 km, the most abundant gas in the atmosphere *is* atomic oxygen, comprising nearly half of all the species present. Nitrogen makes up most of the rest, while life-supporting molecular oxygen is present at only 3%. For reference, the space station is at about 340 km altitude.

Atmospheric gases constitute quite an olio. The following table shows the global mean composition for 2011 (except for nitrogen, oxygen, argon, water, and a few others) as reported* in the IPCC5 (International Panel on Climate Change, Chapter 2 on "Observations: Atmosphere and Surface") with its emphasis on greenhouse gases. The ± precision with which the concentrations are known is excellent.

Species	ppm (parts/million)	Species	ppb (parts/billion)
CO_2	390.48 ± 0.28	SF_6	7.26 ± 0.02
CH_4	1.803 ± 0.005	CF_4	79.0 ± 0.1
N_2O	0.3240 ± 0.0001	C_2F_6	4.16 ± 0.02
		CHF_2CF_3	9.58 ± 0.04
		CH_2FCF_3	62.4 ± 0.3
		CH_3CF_3	12.04 ± 0.07
		CH_3CHF_2	6.4 ± 0.1
		CHF_3	24.0 ± 0.3
		CCl_3F	236.9 ± 0.1
		CCl_2F_2	529.5 ± 0.2
		$CClF_2CCl_2F$	74.29 ± 0.06
		$CHClF_2$	213.4 ± 0.8

Carbon dioxide presently constitutes 0.04% of the atmosphere. That value can also be expressed as 400 parts per million (ppm). There are local variations, indubitably. Natural sources, especially volcanoes, emit about 0.03 atmospheres of carbon dioxide per million years (My), as will be discussed in Chapter 13. To achieve equilibrium if volcanic activity were the only significant source, a steady level of carbon dioxide, 0.03 atm/My, would also have to correspond to the disappearance rate so that there is no net gain or loss (on average). Inverting the value just cited gives an average residence time of carbon dioxide under this extremely simplified alternate point of view. The average residence time, hypothetically, would thus be about 33 million years[†] per atmosphere. (That value is equivalent, under this crude viewpoint, to a "lifetime" of 13,000 years per CO_2 at 400 ppm.)

Contrast the two lifetime estimates with more detailed considerations by the IPCC (the Intergovernmental Panel on Climate Change). The latter group, recognizing that no single lifetime

---

* T. F. Stocker et al., *IPCC, 2013: Climate Change 2013 The Physical Science Basis*, Cambridge University Press, Cambridge, New York (2013).
† 0.03 atmospheres per million years, inverting the ratio gives a million years per 0.03 atm or 33 million years per atm.

can be defined for CO_2 because of the different rates of uptake by the various removal processes, suggests an average lifetime of CO_2 in the air is between 5 and 200 years. Furthermore, the rate at which radiocarbon from atmospheric nuclear tests returned to background levels was mentioned in Chapter 5 as suggesting a mean residence time of about 15 years, reassuringly "within" the given range.

Annual consumption of carbon dioxide by photosynthesis* is currently about 270 gigatons (billion tons, Gt) of carbon dioxide. This produces, reciprocally, 190 Gt of oxygen gas. The total oxygen content of the atmosphere is 1.2 quadrillion tons (1,200,000 Gt). From this, we can estimate that the cycle time for oxygen (dividing the latter value by the former value) is ≈6,000 years (since the total oxygen content has been essentially unchanged for much longer than 6,000 years). What we're saying is, keeping the amount of oxygen in the atmosphere constant, as has been the case more or less for many millennia, the odds of any particular oxygen molecule being removed from the air amount to once every six millennia on the average. A similar approach for the carbon dioxide numbers for photosynthesis, 270 Gt CO_2 removed yet a roughly constant 2,200 Gt CO_2 in the atmosphere suggests an average turnover time of eight years for a photosynthetic carbon dioxide molecule.

CARBON DIOXIDE VARIATIONS

Atop Hawaii, the carbon dioxide concentration at the 11,000 foot tall Mauna Loa peak was measured essentially every month beginning about 1955. This was the first direct measurement of CO_2 in the atmosphere and was the instrumental brainchild of American chemist Charles Keeling of the Scripps Institute of Oceanography.

Charles Keeling

The appearance is now known as the Keeling curve. At this altitude in the islands, the air measured is very clean, monitoring was done continuously, wind corrections were easily accommodated, and instrument calibrations were checked every half hour. The average concentration increased steadily from 315 ppm to just over 400 ppm into the twenty-first century. That's about a 27% change due mostly to increased combustion of fossil fuels worldwide. Superimposed on the gradual increase are very regular and discernible ≈1% annual fluctuations, 12-month cycles, that oscillate from about 3 ppm above the average to 3 ppm below the average. This is illustrated in the figure. The minima are in August; maxima are 6 months apart from minima. Minimum carbon

---

* See Chapter 14.

dioxide consumption and maximum air CO_2 concentrations are in January, winter in Hawaii when plant growth has slowed or even ceased.

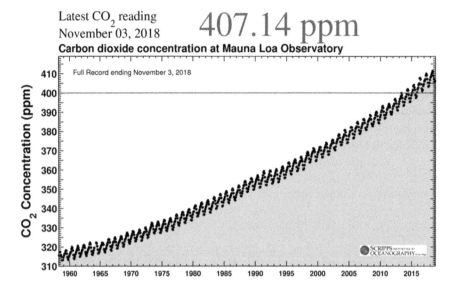

Furthermore, a cycle of the seasonal variations is clearly evident for both the Hawaii location (Mauna Loa) and for studies done in Point Barrow, Alaska. The variations are with respect to the average, shown as zero variation or 0 parts per million on the graph for both the years around 1960 and around 2010.*

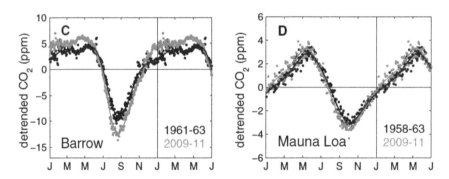

At the South Pole, the same CO_2 measurements were conducted over more than three decades. The results are displayed next (along with data from Samoa in the South Pacific and Point Barrow in Alaska above the Arctic Circle). The average atmospheric CO_2 content again increased over the period studied. Annual oscillations are still clearly manifested, but their magnitude in the South Pole understandably is less by almost a factor of 3 because of the limited amount of photosynthesis associated with the seasons in Antarctica. There, positions of the maxima and minima have shifted one-half year because of season inversion in switching hemispheres from north to south. A careful look at the recent data indicates that although the average CO_2 content was identical in Mauna Loa and the South Pole at the beginning of the studies in the 1950s (not shown), by the 1990s, Mauna Loa is slightly higher. This is consistent with the fact that more fossil fuel is used in the northern hemisphere and with the interpretation that exchange of carbon dioxide between north and south hemispheres is slow compared to the period over which the measurements have been made.

---

* H. D. Graven et al., "Enhanced seasonal exchange of CO_2 by northern ecosystems since 1960," *Science* 341, 1085 (2013).

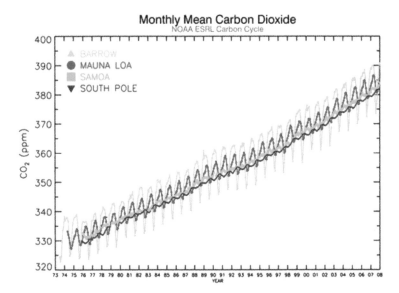

The difference between peak and trough is referred to as the *seasonal amplitude*. There is some more discussion on this in the upcoming chapter on photosynthesis.

All published measurements show averages, a fact that is inherently affected by time resolution. Poor time resolution can conceal useful details. The simplest example is for the Keeling curve figure where a display of annual averages would show a simple, structureless curve rising from about 315 ppm to about 410 ppm, missing all the seasonal amplitudes. Does that mean that monthly averaging reveals everything? Next, we display the 2016 measurements at Mauna Loa taken every hour (gray circles) for one month at the time that the carbon dioxide level had already exceeded the landmark 400 ppm reading in 2013. Note that daily averages (black circles), with even better time resolution than monthly averages, entirely miss the subtly highly variable structure in CO_2 changes that are revealed with the data compiled each hour. Qualitatively, washing out of fluctuations when time resolution is imprecise must be kept in mind when interpreting archived proxies.

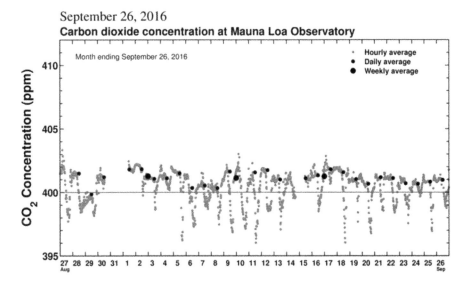

September 26, 2016

HISTORICAL PRESCIENCE ON CARBON DIOXIDE

Back in the 1820s, the French mathematician Joseph Fourier argued that the atmosphere acts like a greenhouse, letting in rays of sunlight, but retaining the rays returning upward from the ground, rays that we now know to be infrared light. His allusion to the greenhouse can actually be traced backward another 18 centuries to the Roman emperor, Tiberius, who had constructed enclosures made from the opaque mineral *mica* that would trap solar heat so that he could quench his craving for cucumbers during the off-season.

In 1894, Swedish geologist Arvid Högbom published some methodical and insightful thoughts on carbon dioxide, or carbonic acid as it was referred to:

> Although it is not possible to obtain exact quantitative expressions for the reactions in nature by which carbonic acid is developed or consumed, nevertheless there are some factors, of which one may get an approximately true estimate, and from which certain conclusions that throw light on the question may be drawn. In the first place, it seems to be of importance to compare the quantity of carbonic acid now present in the air with the quantities that are being transformed. If the former is insignificant in comparison with the latter, then the probability for variations is wholly other than in the opposite case.

> The following calculation is also very instructive for the appreciation of the relation between the quantity of carbonic acid in the air and the quantities that are transformed. The world's present production of coal reaches in round numbers 500 million tons per annum, or 1 ton per $km.^2$ of the earth's surface. Transformed into carbonic acid, this quantity would correspond to about a thousandth part of the carbonic acid in the atmosphere. It represents a layer of limestone of 0.003 millim. thickness over the whole globe, or 1.4 $km.^3$ in cubic measure. This quantity of carbonic acid, which is supplied to the atmosphere chiefly by modern industry, may be regarded as completely compensating the quantity of carbonic acid that is consumed in the formation of limestone (or other mineral carbonates) by the weathering or decomposition of silicates. From the determination of the amounts of dissolved substances, especially carbonates, in a number of rivers in different countries and climates, and of the quantity of water flowing in these rivers and of their drainage-surface compared with the land-surface of the globe, it is estimated that the quantities of dissolved carbonates that are supplied to the ocean in the course of a year reach at most the bulk of 3 $km.^3$. As it is also proved that the rivers the drainage regions of which consist of silicates convey very unimportant quantities of carbonates compared with those that flow through limestone regions, it is permissible to draw the conclusion, which is also strengthened by other reasons, that only an insignificant part of these 3 $km.^3$ of carbonates is formed directly by decomposition of silicates. In other words, only an unimportant part of this quantity of carbonate of lime can be derived from the process of weathering in a year. Even though the number given were on account of inexact or uncertain assumptions erroneous to the extent of 50 per cent. or more, the comparison instituted is of very great interest, as it proves that the most important of all the processes by means of which carbonic acid has been removed from the atmosphere in all times, namely the chemical weathering of siliceous minerals, is of the same order of magnitude as a process of contrary effect, which is caused by the industrial development of our time, and which must be conceived of as being of a temporary nature.

> In comparison with the quantity of carbonic acid which is fixed in limestone (and other carbonates), the carbonic acid of the air vanishes. With regard to the thickness of sedimentary formations and the great part of them that is formed by limestone and other carbonates, it seems not improbable that the total quantity of carbonates would cover the whole earth's surface to a height of hundreds of metres. If we assume 100 metres, – a number that may be inexact in a high degree, but probably is underestimated, – we find that about 25,000 times as much carbonic acid is fixed to lime in the sedimentary formations as exists free in the air. Every molecule of carbonic acid in this mass of limestone has, however, existed in and passed through the atmosphere in the course of time. Although we neglect all other factors which may have influenced the quantity of carbonic acid in the air, this number lends but very slight probability to the hypothesis, that this quantity should in former geological epochs have changed within limits which do not differ much from the present amount. As the process of weathering has consumed quantities of carbonic acid many thousand times greater than the amount now disposable in the air, and as this process from different geographical, climatological and other causes has in all likelihood proceeded with very different intensity at different epochs, the probability of important variations in the quantity of carbonic acid seems to be very great, even if we take into account the compensating

processes which, as we shall see in what follows, are called forth as soon as, for one reason or another, the production or consumption of carbonic acid tends to displace the equilibrium to any considerable degree. One often hears the opinion expressed, that the quantity of carbonic acid in the air ought to have been very much greater formerly than now, and that the diminution should arise from the circumstance that carbonic acid has been taken from the air and stored in the earth's crust in the form of coal and carbonates. In many cases this hypothetical diminution is ascribed only to the formation of coal, whilst the much more important formation of carbonates is wholly overlooked. This whole method of reasoning on a continuous diminution of the carbonic acid in the air loses all foundation in fact, notwithstanding that enormous quantities of carbonic acid in the course of time have been fixed in carbonates, if we consider more closely the processes by means of which carbonic acid has in all times been supplied to the atmosphere. From these we may well conclude that enormous variations have occurred, but not that the variation has always proceeded in the same direction.

Carbonic acid is supplied to the atmosphere by the following processes:--(1) volcanic exhalations and geological phenomena connected therewith; (2) combustion of carbonaceous meteorites in the higher regions of the atmosphere; (3) combustion and decay of organic bodies; (4) decomposition of carbonates; (5) liberation of carbonic acid mechanically enclosed in minerals on the fracture or decomposition. The carbonic acid of the air is consumed chiefly by the following processes: –(6) formation of carbonates from silicates on weathering; and (7) the consumption of carbonic acid by vegetative processes. The ocean, too, plays an important role as a regulator of the quantity of carbonic acid in the air by means of the absorptive power of its water, which gives off carbonic acid as its temperature rises and absorbs it as it cools. The processes named under (4) and (5) are of little significance, so that they may be omitted. So too the processes (3) and (7), for the circulation of matter in the organic world goes on so rapidly that their variations cannot have any sensible influence. From this we must except periods in which great quantities or organisms were stored up in sedimentary formations and thus subtracted from the circulation, or in which such stored-up products were, as now, introduced anew into the circulation. The source of carbonic acid named in (2) is wholly incalculable.

Thus the processes (1), (2), and (6) chiefly remain as balancing each other. As the enormous quantities of carbonic acid (representing a pressure of many atmospheres) that are now fixed in the limestone of the earth's crust cannot be conceived to have existed in the air but as an insignificant fraction of the whole at any one time since organic life appeared on the globe, and since therefore the consumption through weathering and formation of carbonates must have been compensated by means of continuous supply, we must regard volcanic exhalations as the chief source of carbonic acid for the atmosphere.

But this source has not flowed regularly and uniformly. Just as single volcanoes have their periods of variation with alternating relative rest and intense activity, in the same manner the globe as a whole seems in certain geological epochs to have exhibited a more violent and general volcanic activity, whilst other epochs have been marked by a comparative quiescence of the volcanic forces. It seems therefore probable that the quantity of carbonic acid in the air has undergone nearly simultaneous variations, or at least that this factor has had an important influence.

If we pass the above-mentioned processes for consuming and producing carbonic acid under review, we find that they evidently do not stand in such a relation to or dependence on one another that any probability exists for the permanence of an equilibrium of the carbonic acid in the atmosphere. An increase or decrease of the supply continued during geological periods must, although it may not be important, conduce to remarkable alterations of the quantity of carbonic acid in the air, and there is no conceivable hindrance to imagining that this might in certain geological periods have been several times greater, or on the other hand considerably less, than now.

The above paragraphs were originally in Swedish. They were translated into English in 1896 by the renowned Swedish chemist and eventual Nobel laureate Svante Arrhenius. That year, Arrhenius published an article "*On the Influence of Carbonic Acid in the Air upon the Temperature of the Ground.*" Arrhenius was sufficiently impressed with his colleague Högbom's thoughts, that he included the above selections quoted directly into his own discourse. In his article, Arrhenius includes his ideas about the occurrence of glacial periods in relation to carbon dioxide atmospheric variations, something we will examine in Chapter 17.

7 Ye Olde Aire

This goodly frame, the earth, seems to me a sterile promontory; this most excellent canopy, the air, look you, this brave o'erhanging firmament, this majestical roof fretted with golden fire, why, it appeareth no other thing to me than a foul and pestilent congregation of vapours.

W. SHAKESPEARE
(Hamlet)

As part of the carbon dioxide legacy, there is a need to address how to think about the air around us and also how that air composition was constituted over eons. We start with a reminder that early Earth (Chapter 2) was a very turbulent system. Volcanic activity, material bombardment, tectonic creeping of landmasses, continual subductions, churning magma, chemically reactive oceans, and violent weather raining down corrosive chemicals everywhere was pretty much the picture (innocently suggested in the epigraph by Shakespeare), although evidence of the accuracy of this picture of the young planet is extremely indirect.

From the Oxford English Dictionary:

Origin of air (the word) < Anglo-Norman *aeir, aier, eire, eyer, heir, heyr, heyre*, Anglo-Norman and Old French *aire, eir*, Anglo-Norman and Old French, Middle French *aer, air*, Middle French *ayer, ayr* (French *air*) the invisible gaseous substance which envelops the earth and is breathed by all land animals (beginning of the 12th cent. in Anglo-Norman), the atmosphere as a whole (second half of the 12th cent. in Anglo-Norman), contaminated atmosphere, miasma (end of the 12th cent. or earlier), air as one of the four (or more) elements (first half of the 13th cent. or earlier), air in motion, wind, breeze (1275), odour, redolence, the "atmosphere" perceived to be diffused by anything (first half of the 14th cent. or earlier), (in chemistry) any gas or vapour (17th cent.) < classical Latin *āēr* air as a substance, especially as one of the four elements, air, atmosphere, the open air, sky, expanse of air, space, climate, air current, breeze, mist, cloud, odour, scent, in post-classical Latin also breath, spirit (4th cent.) < ancient Greek ἀήρ mist, haze, lower air, air in general, atmosphere, the open air, of uncertain origin.

And ironically the first definition for air in the Oxford English Dictionary:

An atmosphere contaminated by noxious fumes, vapours, etc.; such contaminating fumes themselves; miasma.

For the most part, air is not thought of in such terms today. Back when there was no one to think about air, billions of years ago in fact, the atmosphere was seriously nasty and loaded with carbon dioxide.

MAPPING TIME

For us to discuss history that extends back billions of years, labels prove both convenient and essential. Just as human history uses terms like the "iron age" or the "dark ages" or the "Ming Dynasty," the extensive block of time dating back from the present to about 541 million years ago (Mya) is called the *Phanerozoic* eon, a word whose Greek root is based on *phanero-* for "visible." The label refers to life forms whose remains are large enough and stable enough to have remained visible today: as hard-shelled fossils. The cutoff point for this eon is not arbitrary but corresponds to how far back in time fossils began accumulating. The Phanerozoic Eon is subdivided into three "eras." The most recent is the *Cenozoic,* dating backward from today to 66 Mya. At this specific point of time, a major life extinction event – commonly associated with the disappearance of the dinosaurs – occurred mostly in association with a massive asteroid strike centered on the Yucatan Peninsula of Mexico, but possibly linked with other slightly earlier factors such as eruption of the volcanic Deccan Traps* in India that released vast quantities of CO_2. The name *Cenozoic* derives from the Greek for "new." The preceding era is called the *Mesozoic*, after the Greek for "middle" and carries us back to 251 Mya. The launch of the Mesozoic Era is blamed on massive volcanic eruptions responsible, it is firmly believed, for an earlier round of extinctions of the majority of living species in conjunction with CO_2 release during formation of the extensive Siberian Traps. The still earlier third era is called the *Paleozoic* and extends our nomenclature to 541 Mya. The word's prefix comes from the Greek word for "old" or "ancient." The Cenozoic, Mesozoic, and Paleozoic Eras are further subdivided into periods, epochs, and ages in that order as summarized in the chart.[†] (*Caveat:* Some references scramble the terms eon, era, and period.)

Panel 1

Eonothem/Eon	Erathem/Era	System/Period	Series/Epoch	Stage/Age	numerical age (Ma)
Phanerozoic	Cenozoic	Quaternary	Holocene		present / 0.0117
			Pleistocene	Upper	0.126
				Middle	0.781
				Calabrian	1.80
				Gelasian	2.58
		Neogene	Pliocene	Piacenzian	3.600
				Zanclean	5.333
			Miocene	Messinian	7.246
				Tortonian	11.63
				Serravallian	13.82
				Langhian	15.97
				Burdigalian	20.44
				Aquitanian	23.03
		Paleogene	Oligocene	Chattian	28.1
				Rupelian	33.9
			Eocene	Priabonian	37.8
				Bartonian	41.2
				Lutetian	47.8
				Ypresian	56.0
			Paleocene	Thanetian	59.2
				Selandian	61.6
				Danian	66.0
Phanerozoic	Mesozoic	Cretaceous	Upper	Maastrichtian	72.1 ±0.2
				Campanian	83.6 ±0.2
				Santonian	86.3 ±0.5
				Coniacian	89.8 ±0.3
				Turonian	93.9
				Cenomanian	100.5
			Lower	Albian	~113.0
				Aptian	~125.0
				Barremian	~129.4
				Hauterivian	~132.9
				Valanginian	~139.8
				Berriasian	~145.0

Panel 2

Eonothem/Eon	Erathem/Era	System/Period	Series/Epoch	Stage/Age	numerical age (Ma)
	Mesozoic	Jurassic	Upper	Tithonian	~145.0
				Kimmeridgian	152.1 ±0.9
				Oxfordian	157.3 ±1.0
			Middle	Callovian	163.5 ±1.0
				Bathonian	166.1 ±1.2
				Bajocian	168.3 ±1.3
				Aalenian	170.3 ±1.4
			Lower	Toarcian	174.1 ±1.0
				Pliensbachian	182.7 ±0.7
				Sinemurian	190.8 ±1.0
				Hettangian	199.3 ±0.3
					201.3 ±0.2
		Triassic	Upper	Rhaetian	~208.5
				Norian	~227
				Carnian	~237
			Middle	Ladinian	~242
				Anisian	247.2
Phanerozoic			Lower	Olenekian	251.2
				Induan	252.17 ±0.06
	Paleozoic	Permian	Lopingian	Changhsingian	254.14 ±0.07
				Wuchiapingian	259.8 ±0.4
			Guadalupian	Capitanian	265.1 ±0.4
				Wordian	268.8 ±0.5
				Roadian	272.3 ±0.5
			Cisuralian	Kungurian	283.5 ±0.6
				Artinskian	290.1 ±0.26
				Sakmarian	295.0 ±0.18
				Asselian	298.9 ±0.15
		Carboniferous / Pennsylvanian	Upper	Gzhelian	303.7 ±0.1
				Kasimovian	307.0 ±0.1
			Middle	Moscovian	315.2 ±0.2
			Lower	Bashkirian	323.2 ±0.4
		Mississippian	Upper	Serpukhovian	330.9 ±0.2
			Middle	Visean	346.7 ±0.4
			Lower	Tournaisian	358.9 ±0.4

Panel 3

Eonothem/Eon	Erathem/Era	System/Period	Series/Epoch	Stage/Age	numerical age (Ma)
					358.9 ±0.4
		Devonian	Upper	Famennian	372.2 ±1.6
				Frasnian	382.7 ±1.6
			Middle	Givetian	387.7 ±0.8
				Eifelian	393.3 ±1.2
			Lower	Emsian	407.6 ±2.6
				Pragian	410.8 ±2.8
				Lochkovian	419.2 ±3.2
		Silurian	Pridoli		423.0 ±2.3
			Ludlow	Ludfordian	425.6 ±0.9
				Gorstian	427.4 ±0.5
			Wenlock	Homerian	430.5 ±0.7
				Sheinwoodian	433.4 ±0.8
			Llandovery	Telychian	438.5 ±1.1
				Aeronian	440.8 ±1.2
				Rhuddanian	443.8 ±1.5
Phanerozoic	Paleozoic	Ordovician	Upper	Hirnantian	445.2 ±1.4
				Katian	453.0 ±0.7
				Sandbian	458.4 ±0.9
			Middle	Darriwilian	467.3 ±1.1
				Dapingian	470.0 ±1.4
			Lower	Floian	477.7 ±1.4
				Tremadocian	485.4 ±1.9
		Cambrian	Furongian	Stage 10	~489.5
				Jiangshanian	~494
				Paibian	~497
			Series 3	Guzhangian	~500.5
				Drumian	~504.5
				Stage 5	~509
			Series 2	Stage 4	~514
				Stage 3	~521
			Terreneuvian	Stage 2	~529
				Fortunian	541.0 ±1.0

The Phanerozoic is preceded by the ultra-eon, the Precambrian, dating back over 4 billion years. The Precambrian span's infancy is preceded by the Hadean Eon (to 4,600 Mya). The name (from the Greek "Hades") is in reference to the hellish conditions, heat in particular, of the times. Numerical model calculations imply that high temperatures caused very high rates of carbon dioxide outgassing during the Hadean, building up ambient pressures of at least one atmosphere of CO_2 and perhaps as much as ten atmospheres compared to today's 0.0004 atmospheres. The greenhouse effect would have kept surface temperatures warm at these times, times which are known to have

---

* See Chapter 13.
† From the official 2015 version of the chart of the International Commission on Stratigraphy.

experienced lower solar radiation, a situation that otherwise would have been able to sustain quite cold climates. We have limited knowledge of the atmosphere so long ago, but it is nevertheless surprising how confident we are in describing it, inferring details from the clever use of proxies. (See the next chapter.) Some experts studying characteristics of craters on the lunar surface, for example, suggest that there was heating of Earth from over a dozen impacts during this eon by asteroid-like objects measuring more than 100 miles in width.

HISTORY OF THE ATMOSPHERE

A very broad stroke picture of developments in the atmosphere over its first few billion years of existence has the carbon dioxide level plummeting maybe as much as 1,000-fold due mostly to weathering (chemical reactions with rock, see Chapter 16). The oxygen concentration slowly burgeoned 10,000-fold, mostly due to photosynthesis (see Chapter 14). Some relevant highlights in this history germane to the legacy of carbon dioxide follow.

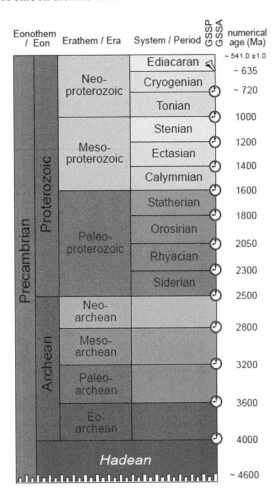

PRECAMBRIAN AIR

There are few arguments against the idea that carbon dioxide levels were high eons ago. Among the simplest arguments supporting the point of view is to look at the atmospheres of Venus and Mars. Our two sibling planets, Venus and Mars, chaperone the iconoclast Earth, yet probably have never been modified themselves by biogeochemical influences. Picking a midpoint of the Venus/ Mars atmospheric conditions gives a naive indication of how the atmosphere would have been in the absence of life on Earth: mostly carbon dioxide and of the order of one atmosphere pressure. (See Chapter 2.)

Evidence for such high concentrations of carbon dioxide billions of years ago is flimsy though. But the need for such an atmosphere is strongly supported by even the roughest approximations to global temperature requiring a greenhouse effect to maintain the planet above freezing despite per- haps 30% less solar radiation. Moreover, profuse quantities of limestone and other mineral deposits derived from atmospheric CO_2 easily hint at an atmosphere loaded with the gas. Carbon dioxide likely arose from and was continuously fed by prolific volcanic activity coupled with absence of effective CO_2 removal venues at the time. The Precambrian stage covers nearly 90% of Earth's 4.6-billion-year existence yet clear records are scarce. A critical review of competing scenarios is given by Lyons et al.*

In his book *Molecular Mechanisms of Photosynthesis*, Blankenship summarizes what is known about the rate at which the level of carbon dioxide dropped a thousand-fold since the Precambrian. The trend is sketched in the figure, along with estimates of the overall accompanying growth of oxygen, very likely as the result of photosynthesis during the latest 2 billion years plus. Somewhere between 2.4 and 2.0 billion years ago, oxygen jumped from mere parts per million in the atmosphere to perhaps 0.02 atmospheres, one-tenth the current value in a phenomenon known as the *Great Oxidation Event*. These estimates and their timing remain very, very uncertain though. For the next billion years, evidence implies the level of atmospheric oxygen remained roughly unchanged.

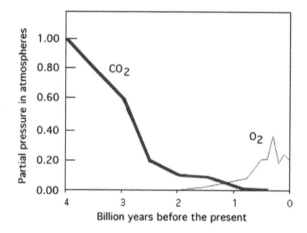

Chemically reactive, oxygen can be toxic, a characterization that seems inconceivable at first. Oxygen is chemically quite aggressive, having the ability to be transformed into the even more egre- gious peroxide and superoxide forms. Early life forms seem not to have evolved the ability to keep oxygen's destructive reactivity in check. After the origin of photosynthetic species (marine bacteria around 3 billion years ago), the by-product oxygen in the oceans (and, by exchange, in the atmo- sphere) was kept low by its reacting with iron and other active chemicals dissolved in the ocean or by microbial metabolic consumption of dead organic matter. However, this *status quo* was to change dramatically. Among the proposed explanations is the possibility of tectonic shifts in continental shelves which allowed the dead organic matter to be buried before being thoroughly metabolized, consequently preventing consumption of oxygen. Also, microbes that had produced methane as a by-product required the element nickel (element 28) as a micronutrient to enable its biochemistry.

---

* T. W. Lyons, C. T. Reinhard, and N. J. Planavsky, "The rise of oxygen in Earth's early ocean and atmosphere," *Nature* 506, 307–315 (2014).

But the level of nickel in the oceans, probably supplied by volcanic activity, petered out. Until this reversal, methane also consumed appreciable quantities of life-threatening atmospheric oxygen by reacting with it to produce carbon dioxide, in essence, a very slow combustion process.

The late Precambrian also witnesses severe, extended frigid periods known as "snowball Earth" events (see Chapter 17). These were so influential that one division of time in the Precambrian span is named the Cryogenian Period (from *cryos*, Greek for icy cold).

The discovery of dozens of exoplanets in recent years has prompted schemes for recognizing the existence of alien life. Among these approaches is one that taps the intimate connection between atmospheric oxygen and photosynthetic use of carbon dioxide. The O_2–CO_2 link can be explored spectroscopically from afar.*

PHANEROZOIC

The 2007 Intergovernmental Panel on Climate Change (IPCC) published a compilation of atmospheric carbon dioxide levels interpreted as representing changes over the most recent half-billion years, the Phanerozoic Eon, which proceeds through the Paleozoic, Mesozoic, and Cenozoic Eras. Measurement methods employed were based on a combination of *isotopic fractionation* techniques in which the relative amounts – ratios – of isotopes of a particular element (such as $^{13}C/^{12}C$, $^{18}O/^{16}O$, $^{10}B/^{11}B$, or for strontium isotopes, $^{87}Sr/^{86}Sr$) change in understandable ways as chemical and/or physical conditions are altered and also on the physical appearance of leaf fossils themselves. We discuss the very informative isotope fractionation technique's use in more detail in Chapter 20. Other techniques involved the use of *stomatal indices* based on leaf pores. Various proxy methods like these are looked at in more detail in the next chapter. Results in the IPCC figure did not agree well when various proxy measurements were compared. Those serious discrepancies have since been largely settled in 2011 by a more thorough consideration of how carbonates in soils reflect atmospheric carbon dioxide levels.

Known ice ages in the Phanerozoic Era are indicated in the figure (from 2018[†]). The bars dropping from the top border represent the latitude extent (left axis) of continental ice. The line through the gray-shaded region is the most likely fit to the carbon dioxide data taken from a variety of proxy measurements. The slowly decreasing curve is a linear fit to the data. Carbon dioxide values are indicated on the right axis and the faint dashed horizontal line represents the preindustrial level of CO_2.

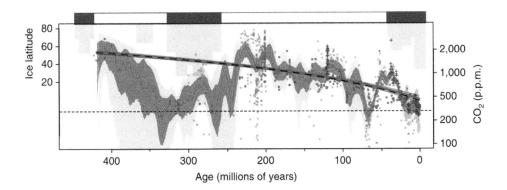

* V. S. Meadows et al., "Exoplanet biosignatures: Understanding oxygen as a biosignature in the context of its environment," *Astrobio.* 18, 630–662 (2018).
† G. L. Foster, D. L. Royer, and D. J. Hunt, "Future climate forcing potentially without precedent in the last 420 million years," *Nat. Commun.* 8, 14845 (2017).

EARLY PALEOZOIC ERA

An interesting aspect of the early Paleozoic Era is the finding of fossil charcoal residues. These indicate that wildfires became possible about 420 Mya when terrestrial plants evolved and atmospheric oxygen surpassed 13%. Dead plants can become fuel, especially coal and organic-rich shales. The Marcellus shale formation in the U.S. Appalachian mountain region has been dated to about 390 Mya. Before this time, there was essentially no grass, no brush, nor any trees to burn. Marine plants do not form coal.

A dip in CO_2 centered around 300 Mya is associated with the Carboniferous Period. At that time, it can be reasonably argued that the rise of large plants would have increased photosynthesis activity and the weathering of ground and soil silicates. Photosynthesis and weathering extract carbon dioxide from the atmosphere. At the left of the figure, high carbon dioxide levels (extending back to before ~500 Mya) would arguably be consistent with warming and the end of the earlier *snowball Earth* (a hypothesis explained in more detail in Chapter 17), and would have been generated as the result of volcanic activity not counterbalanced by carbon dioxide–consuming processes.

CARBONIFEROUS PERIOD

During the Carboniferous Period, (roughly 360–300 Mya), atmospheric oxygen levels rose to as high as 35%, probably owing to the evolution and proliferation of lignin-incorporating terrestrial plants. Biomaterials such as lignin whose complex chemical structure is shown in Chapter 14, are resistant to oxygen-consuming decomposition processes. In the Carboniferous, lignins became prevalent, even more so than in present day plants. There is some thought also that microbes had not evolved yet to decompose wood. Dead plants would then accumulate. Lignin, a major component of wood and other plant materials, was being produced copiously from carbon dioxide by newly evolving vascular plants such as tall seed ferns characteristic of this period. However, a 35% oxygen peak level would have been stressful for photosynthetic systems since high oxygen boosts *photorespiration*. Photorespiration, a process in which oxygen is consumed metabolically by plants, is the reverse of photosynthesis. As the prefix "photo-" indicates, photorespiration is triggered by light. Photorespiration reduces the efficiency of photosynthesis in that carbon dioxide is released rather than harvested. Photorespiration is discussed in more detail in Chapter 15.

High oxygen in the Carboniferous Period is also concurrent with the evolution of giant insects like a millipede nearly as long as an alligator, and dragonflies as large as falcons. Consumption of oxygen to fuel metabolism is substantial for flying species even today. The period of these ancient superflyers' existence matches the hypothetical elevated oxygen content in the air. This picture of the Carboniferous Period is consistent with the formation of coal deposits, deposits being generated in vast quantities one way or another from CO_2. At stages during the Carboniferous, CO_2 reached 10 times today's level. Nevertheless, from about 326 Mya and lasting for some 60 million years, the longest and most extensive glaciation of the Phanerozoic Eon occurred in sympathy, so to speak, with a drastic drop in atmospheric carbon dioxide levels. This is seen as a valley (bottoming at less than 500 ppm) in the smooth curves, representing averages, of the previous figure. Higher CO_2 levels are explained by interspersed phases between glaciations shown in the figure, rather warm interludes.

Why didn't oxygen percentages continue to climb? That is a very logical question. There are upper constraints on the levels of oxygen in the atmosphere. We are at 21% now. At oxygen less than 10% atmospheric pressure, large animals will not survive. That is not a constraint but rather a reflection that large animals require higher availability of oxygen. In contrast, at 30%, wildfires would be global, effectively capping the degree of oxygen enrichment. During the late part of the Carboniferous and subsequently into the early Permian Periods at the end of the Paleozoic Era, overwhelming evidence suggests that oxygen levels were at their upper limit. (See the dotted curve in the figure.*) One idea raised about the atmosphere then is that the total amount of atmospheric nitrogen was constant. That currently amounts to 0.78 atmospheres. When oxygen increased, the total atmospheric pressure therefore increased as well. This means the density of air increased; buoyancy of giant insects thus becomes easier to explain. However, 251 Mya, at the end of the Permian Period, the last in the Paleozoic Era, oxygen had fallen to 15%, coincident with the most effective mass extinction ever although disentangling cause and effect is complex. Fossil signs of giant insects completely disappeared when oxygen levels bottomed, effectively asphyxiating the gargantuan, oxygen-gulping bugs. Thus ended the Paleozoic Era.

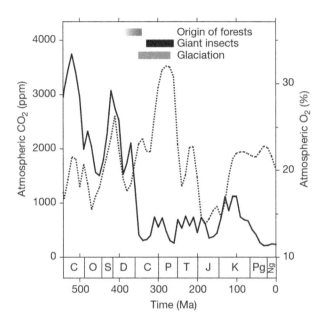

In the figure, C = Cambrian, O = Ordovician, S = Silurian, D = Devonian, C = Carboniferous, P = Permian, T = Triassic, J = Jurassic, K = Cretaceous, Pg = Paleogene, and Ng = Neogene Periods.

MESOZOIC ERA

About 250 Mya, carbon dioxide levels in the atmosphere rose to greater than 1,000 ppm launching a general extended greenhouse phase of warm climate with CO_2 rising to as much as 6,000 ppm, punctuated occasionally by cool, but not glacial, episodes. A more detailed look is graphed here for the fluctuating carbon dioxide levels over the Jurassic and subsequent Cretaceous Periods (K) in the Mesozoic Era obtained by several different proxy methods: paleosols refer to soil carbonates, stomata are leaf anatomical structures, phytoplankton are marine algae, and liverworts provide fossilized leaves.† Over 700 data points were averaged into 10-million-year bins for the figure.

* D. L. Royer, "Atmospheric CO_2 and O_2 during the phanerozoic: Tools, patterns and impacts," *Treatise on Geochemistry,* 2nd edition, Chapter 6, 251–267 (2014).
† D. L. Royer, op. cit.

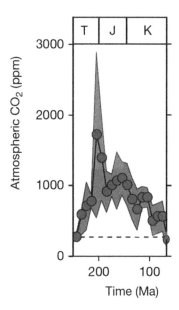

The Cretaceous Period (145–65 Mya), the most recent stretch of the Mesozoic Era, was character-ized by massive growth of oceanic crust and extensive rifting – tearing apart of Earth's crust – prob-ably accompanied by undersea volcanic release of carbon dioxide leading to greenhouse warming. During this period, South America, Antarctica, and Australia broke away from the African land-mass. Large dinosaurs and marine reptiles flourished. Small mammals and birds appeared as did flowering plants. Results of various studies indicate carbon dioxide levels above 1,000 ppm. Excursions to much lower levels occurred intermittently. Extreme Arctic sediments dating to that period contain not only plant fossils consistent with warm climate near the pole and absent in the current frigid climate, but even hold crocodile-like fossils, certainly convincing of a warm climate.

As proxies from ever younger periods are analyzed, the carbon dioxide picture becomes less uncertain, its legacy coming into better focus as we will see.

CENOZOIC ERA

Sixty-six Mya, a cataclysm ended the Cretaceous Period and the Mesozoic Era. Triggered by the impact of a massive asteroid on the Yucatan Peninsula in Mexico, a landmass composed mostly of limestone, the ensuing devastation is most noteworthy in the public eye for its popular association with the extinction of the dinosaurs. This event is perhaps the most exhaustively investigated of all geological episodes. The evidence is extraordinarily convincing. It includes a 100-mile diameter crater discovered in the 1970s and the finding of material ejected by the impact radiating out from the colli-sion location. Ejected material with composition specific to the impact site occurs with decreasing fre-quency yet as far away as thousands of kilometers. Over 350 sites have been investigated worldwide.[*] Physics calculations have been used to hypothesize that the collision was forceful enough to have generated earthquakes exceeding a Richter magnitude of 11. Carbonates abundant at the impact site would have been thermally vaporized, decomposing into voluminous quantities of CO_2. Examination of fossils in layered deposits dated to this time show sudden extinction of the majority of species fol-lowed by the appearance "shortly afterward" of other species. (See the next figure.[†]) The emergence

---

[*] P. Schulte et al., "The Chicxulub asteroid impact and mass extinction at the Cretaceous-Paleogene boundary," *Science* 327, 1214–1218 (2010).

[†] C. Lécuyer, "Learning from past climate changes," *Science* 360, 1400–1401 (2018).

of new populations and also the appearance of newly evolved species would be reasonably expected because of vanished predators or purged competitors.

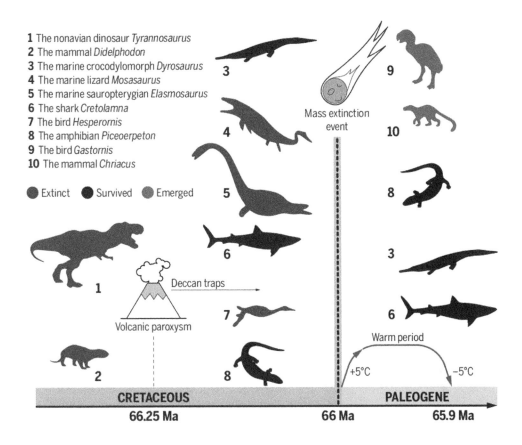

One study of a core drilled from the North Atlantic, displayed here, reveals a precipitous drop in ^{13}C isotope fractions in deposits. Successive data points are "mere" thousands of years apart. Besides being triggered by an abrupt event of some kind, the detailed interpretation of that event can have multiple explanations associated with the asteroid impact and its immediate after effects. Also found, the abundance of calcite (carbonate shells and skeletons) in the deposits drops abruptly from 75% to less than 20% in the top part of the figure suggesting disappearance of marine species and/or increased ocean acidity (from increased atmospheric carbon dioxide levels) causing carbonates to dissolve. (See Chapter 12.)

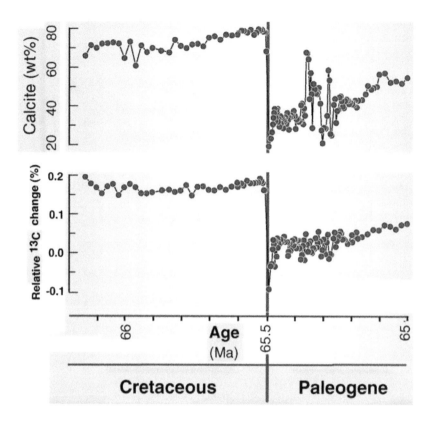

The most recent 66 million years, the Cenozoic Era beginning with the Paleogene Period, followed the Yucatan devastation. The era is characterized by global cooling with numerous, extensive glacial periods, especially over the last half-million years. The era has been explored extensively by multiple experimental techniques. Various results are displayed in two graphs.

The first graph represents enrichment of the rare ^{18}O isotope in the calcium carbonate shells (from CO_2) of Cenozoic Era foraminifera, submillimeter protozoa. The results were obtained from hundreds of deep sea drilling projects. Interpreting the ^{18}O signal is enormously complex with many factors contributing in competing ways. But to start with a simple perspective, the growth of carbonate shells in ocean water can directly reflect the ^{18}O enrichment of the sea. Melting ice distorts the degree of enrichment. Here's how. Glaciers grow from accumulating snow layers known to be enriched in the *light* isotope of oxygen. The reason is that light oxygen prevails in evaporation from the sea – it's more easily lost – depleting oceans of the ^{16}O isotope. (See Chapter 20.) Accordingly, the presence of high ^{18}O enrichment in shells accreted in the ocean waters during glacial periods because a portion of the lighter isotope has left. On the other hand, when glaciers melt *en masse*, the enrichment of the heavier isotope in the oceans is diluted by the influx of large quantities of the low-^{18}O fresh water, water that has accrued the lighter ^{16}O from melting snow. Relative enrichment differences can be determined very precisely even though the actual values change by only minute amounts. Interglacial period meltwaters would then be characterized by reduced ^{18}O levels in foraminiferal $CaCO_3$. Oxygen isotope fractionations, if all other effects are arguably negligible, are proxies for the volume of water tied up in glaciers; that is, they indicate ice volume. Extensive glaciation is shown on the figure by broad horizontal bars "growing" toward the right border of the figure.

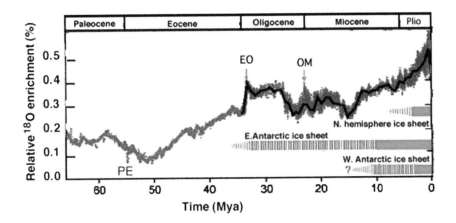

Continuing with an interpretation of the figure, nearly 55 Mya, the low relative enrichment of heavy oxygen on the left of the figure suggests there was not much ice volume, glaciers having melted and diluted the ocean's fraction of ^{18}O. A sharp thermal maximum at this time, at the Paleocene–Eocene crossover (PE in the figure), is supported by numerous other pieces of evidence and suggests that the carbon dioxide component of the atmosphere was about 0.2% (2,000 ppm, see later figure), some five times higher than at present. Temperatures were warmer by an additional 5–7°C. With the passage of time, cooling is implied by the rise in the ^{18}O enrichment of foraminiferal shells shown here.

By the Eocene–Oligocene (EO) "boundary" at about 34 Mya, a spike in enrichment is correlated very closely with a sudden onset of Antarctic ice sheets (horizontal band). Antarctic ice has been around continuously since then. The next epochal boundary shows less dramatic evidence of a deep chill at about 23 Mya, marking the Oligocene–Miocene transition (OM), again connected with glacial maxima and the extinction of cold-intolerant species as evidenced by dated fossil identifications. Only within the last 2–3 million years did the *Northern* hemisphere ice sheets appear, most dramatically in Greenland. Despite the superb fractionation measurements, the interpretation in terms of ice volume is confounded by the additional influences of ocean acidity and temperature on ^{18}O enrichment shown in the figure.

In addition to ^{18}O/^{16}O fractionation determinations, ^{13}C/^{12}C isotope ratios have been measured in those foraminiferal calcium carbonates and are shown here. But interpretation of the carbon fractionation results is a frustratingly complex exercise.

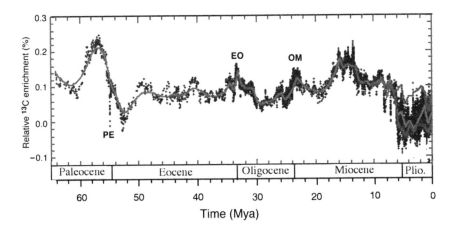

Finally, results for atmospheric CO_2 levels are next displayed. The figure is from Figure 5.2 of the Fifth Intergovernmental Panel on Climate Change (*q.v.*). In the figure, the uncertainty is indicated by the broad gray band. Problematic is the observation that variations by as much as a factor of 2 derived from different proxy techniques still cast uncertainty overestimates of carbon dioxide in ancient times.* The upper dashed horizontal line at 400 ppm is the present level of carbon dioxide. The lower dashed line is the preindustrial level. Extracting carbon dioxide levels from these various archives of proxies is obviously not as precise as the isotope fractionation measurements displayed here. The ability to pinpoint time, as can be seen in the horizontal lines through the individual data points, is not too good, typically millions of years. Dramatic spikes, up or down, seen for the oxygen and carbon isotope fractionations at the Paleocene–Eocene (PE) boundary for example, are not readily apparent in the CO_2 results extruded from the various proxy studies.

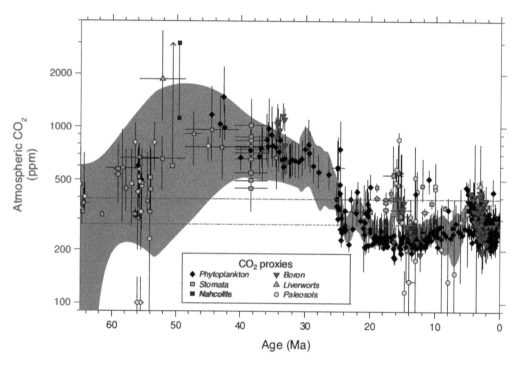

But a very general trend emerges, showing the carbon dioxide level dropping approximately continuously starting around 50 Mya. Among the contributing factors to the drawdown of carbon dioxide from the atmosphere, the following two geophysical phenomena have been reasonably proposed. One is decreased volcanic contribution of carbon dioxide input to the atmosphere owing to a lessening of the rifting in the North Atlantic. The second is the exposure of massive tectonic uplifts with the birth of the Himalayas and their additional contribution to pulling down carbon dioxide by means of weathering of silicate bearing rocks. (See Chapter 16.)

The regrettable scattering of the various proxy results for atmospheric carbon dioxide levels in the previous figure should be compared with what can be observed for the most recent million years. Direct measurement of CO_2, trapped in ice, is part of the discussion in the next chapter, one that describes some proxies tracing recent behavior in more critical detail.

---

* D. J. Beerling and D. L. Royer, "Convergent cenozoic CO_2 history," *Nat. Geosci.* 4, 418–419 (2011).

8 Proxies

What we see depends mainly on what we look for.

SIR JOHN LUBBOCK
British biologist and archaeologist

There are no facts, only interpretations.

FRIEDRICH NIETZSCHE
German philosopher

Almost all of the ideas/hypotheses from the previous and subsequent chapters are predicated on the use of proxies. We need to provide some detailed explanations and examples to foster at least a partial appreciation for the role that proxies play in exploring the legacy of carbon dioxide. Dictionaries define a proxy as something or someone authorized to act for another. Conceptually, the proxy is a surrogate, a substitute representative. Some definitions refer to an authorized role, acting on behalf of another. Although such stringency would be a welcome situation, it is an overstatement. The idea is to find something measurable that "co-varies" with a feature of interest. Sulfur deposits are proxies for volcanoes for instance. Economists use the consumer purchasing index (CPI) as a proxy for inflation and the real value of wages. Physicians use blood creatinine levels as a diagnostic tool for kidney problems. Ecologists use penguin counts as a proxy for krill abundance in Antarctic waters. They even back up further and use distant measurements of penguin droppings as proxies for the penguin counts. Atmospheric carbon dioxide content over the ages serves in this proxy capacity. Average temperatures implied by average carbon dioxide levels should be supported by experiments accompanied by sound arguments as to why the connection – the co-varying – would be applicable in the distant past as well. The co-variance, using carbon dioxide as an example, could be something influenced by changes in CO_2, or something whose own variations influence CO_2, or some third phenomenon that influences both. Correlation could be cause, or effect, or both (feedback, that is), or accidental. Ideally, no other phenomena would affect the interrelationship. That is rarely the situation. The use of proxies invariably involves adjustments to accommodate competing influences or interferences. Confidence in any interpretation is much enhanced by the use of multiple proxy methods.

Proxies include such substances as dust, pollen, ash, soot, methane, and an extensive variety of chemicals as simple as elemental beryllium (element 4), or complex organic structures consisting of dozens of atoms. Sources of proxy materials include sea beds, lake sediments, tree rings, cave deposits, ice cores, and rock layers to cite some examples. Measurements entail isotope abundance determinations, gas compositional analyses, microscopic examination, conductivity measurement of electric current, and radioactive decay. The proxies, their sources, and quantitative and qualitative measurement techniques are all founded on strong scientific principles. The fourth aspect is problematic: interpretation.

Promising proxies are numerous. New possibilities arise all the time. Only a few examples will be discussed in depth so as to convey their utility and drawbacks.

BORATE PROXY

A clever proxy for carbon dioxide levels is the small amount of the aqueous ion *borate*, $B(OH)_4^-$, containing element 5, boron (B). Boron is one of the rarer elements in Earth's crust and the solar system and even in the universe. Its abundance is 0.001% of the crust. For comparison, silicon (Si, element 14) is 28% of the crustal composition. In nature, boron as a trace element plays a role in strengthening plant cell walls. Small amounts of borate get incorporated into the carbonate shells of the marine protozoan *foraminifera*. The shells of dead foraminifera accumulate as sediments. The sediments can be dated with some confidence and their borate content can be studied. The boron in borate has two stable isotopes, ^{10}B and ^{11}B, for which the isotopic fractionation (see Chapter 20) effect on accumulation in shells can be quite dramatic. The extent of fractionation, how much the chemistry differs for the two masses of boron, has been thoroughly studied and is very dependent upon pH at acidities representative of the marine environment, that is, around pH 8, because of the equilibrium reaction with *boric acid*, $B(OH)_3$.

$$B(OH)_3 + H_2O \rightleftarrows B(OH)_4^- + H^+$$

The pH dependence of isotopic mass fractionation can be calibrated. Consequently, $^{11}B/^{10}B$ abundance measurements in carbonate sediments can be used to yield, in the proxy sense, pH values of the ocean at the time of sedimentation, or more accurately, at carbonate shell formation. The pH, in turn, is determined mostly by the carbon dioxide content of the atmosphere in equilibrium with ocean surfaces where foraminifera grow. The lower the pH implied, the higher the atmospheric carbon dioxide inferred and *vice versa*. (See Chapter 10.)

Among the few difficulties encountered applying the borate proxy to expose atmospheric carbon dioxide information is the assumption that contributions to the isotope fractions from river runoffs have been constant. That has proven not to be the case in recent experimental studies and the degree to which this problem affects interpretations is under discussion. Part of the complication is that river waters have been observed to encompass a wide range of pH, undermining the supposition of constancy.

READING ABOUT THE ATMOSPHERE IN ICE

When they first accumulate, typical snowfalls are fluffy and have densities in the range of 10–30% of that of liquid water. Most ski buffs hope for accumulations of many feet, not uncommon in certain regions of the world. At depths reaching 15 m below the top surface of accumulations, the snow packs down and reaches densities of more than half that of liquid water. Air is still mixed in with the snow at this depth. If there are no warm seasons to melt these accumulations, the build-ups continue. The deeper snow compacts even further and is typically referred to as *firn*, a word of German origin referring to "last year." At depths of 50 m, the density has increased to 80% that of water due to the increased weight above. Air can still circulate slightly because of persisting, but ever diminishing, porosity. Eventually, below a so-called close-off depth – usually 60 to 100 m – air becomes entombed. Trapped air is somewhat younger than the snow around it because circulation was occurring before sealing. Below 100 m in depth, ice forms, but at densities lower than normal ice due to trapped and now isolated air.

A May 2010 *Wall Street Journal* feature article, "Mining for Cold, Hard Facts," highlighted a recent expedition to Antarctica that was drilling the ice core for analysis 1000 km from the South Pole where there is unusually high snowfall. A staff of 45 persons was necessary for this very complex operation. Part of the motivation was to determine the answer to a very important question: whether increases in gases such as carbon dioxide preceded or followed temperature increases in the past. Ice core studies appear to be the only way to resolve that question, barring the invention of a true time machine.

Archived in the layers of ice are multiple proxies for what was in the air at the time, including not only trace gases such as carbon dioxide and methane but also dozens of substances containing a variety of elements. Accessing these proxies at various times from the past is as "easy" as drilling out a core from the ice and performing chemical and physical analyses.

The age of core samples found relatively shallow can be dated visually. Annual layers can be seen as color variations due mostly to seasonal variations: dust, wind-driven sea salt, and the like. Relatively recent layers can be cross checked with dates of documented volcanic eruptions. Even deeper cores frequently display clear layering. is the figure shows a 1 m long section of an ice core raised from over 1,800 m depth (perhaps a half-million years old) in Greenland. The tiny tick marks at the bottom measure off centimeters increasing from left to right.

Deeper down, ice has compressed and the layers become increasingly difficult to distinguish. Different dating methods often give different results. The discordance increases with depth. Disparities can amount to several thousand years at 2 km depth. However, bear in mind that such uncertainty will amount to only a couple of percent of the age itself. An excellent description of these studies and the science behind them can be found in Richard Alley's *The Two-Mile Time Machine*. Similarly, the investigations in Greenland and also in Antarctica have been described in the very readable *The Ice Chronicles* by Paul Andrew Mayewski and Frank White.

There is a previously mentioned aspect to the dating issue. Even though the ice layers themselves can be assigned to an age with a good degree of precision, the trapped carbon dioxide has spread within the layers, diffusing and mixing with older and younger gas up until a seal forms. The less accumulation there has been of snowfall, the greater the uncertainty of the age of the trapped gas. Trapped gas is always younger – as much as 400 years – than the confining ice.

In Central Antarctica, the low temperatures and modest snowfall accumulations limit the exchange of gases at different depths and provide samples that date back thousands of years. Analysis of ice cores from Antarctica by a consortium of European laboratories gives us an inkling about ancient carbon dioxide concentrations in the atmosphere and, indirectly, temperature variations. Scientists reconstructed the history of carbon dioxide (measured from bubbles in ice cores) over the past 800,000 years.[*] Cores were retrieved from depths as great as 3,200 m. At such depths, the estimated time interval covered by each core section sampled and analyzed was 570 years. (Any structure within those 570 years would be washed out and averaged.) Measurements were shared among different laboratories using several different carbon dioxide extraction techniques and various composition analyses in order to ensure confidence in the results. In the figure, approximate times at which various glacial periods terminated are indicated by "T" with Roman numeral subscripts. The record low for CO_2 is 171.6 ± 1.4 ppm.[†] Ice-core carbon dioxide is not a proxy for carbon dioxide. It is as close to a direct measurement as one can get without an actual time machine. However, the carbon dioxide level can itself serve as a proxy for other quantities of interest, temperature being one.

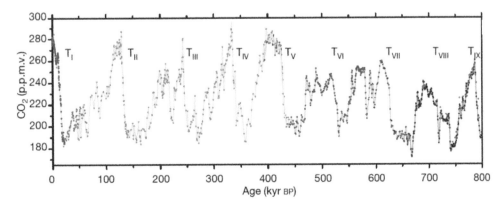

[*] Drilling is expected to start in 2020 in Antarctica to extend the reach to 1.5 million years. Q. Schiermeier, "Speedy Antarctic drills start hunt for Earth's oldest ice," *Nature* 540, 18–19 (2016).

[†] B. Bereiter et al., "Revision of the EPICA dome C CO_2 record from 800 to 600 kyr before present," *Geophys. Res. Lett.* 42, 542–549 (2015).

The same ice core samples were also analyzed for hydrogen isotope fractionation in the ice's water itself. Heavy hydrogen – deuterium – is discriminated against when ocean water vaporizes later to fall as snow. The degree of fractionation is extremely sensitive to temperature (see Chapter 20) because the heavy-to-light mass ratio on which the phenomenon is based, hydrogen (^{2}H:^{1}H masses) is 2, compared to less than 1.1 (10%) for both carbon and oxygen isotope pairs. Next is a comparison of the deuterium fractionation measurements in ice displayed with the carbon dioxide content of ice-trapped air (identical to the previous figure) showing how precisely they correspond in behavior.

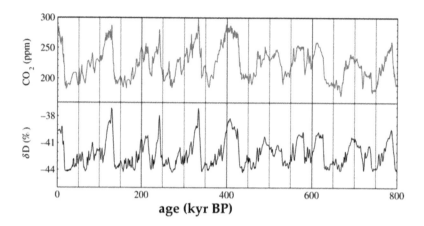

If the recovered cores from these 800 millennia extractions were laid down end to end in a continuous assembly, the length would exceed that of the world's longest bridge, the Akashi Strait suspension bridge in Japan, shown here just to give an idea of how extensive a 3 km ice cylinder would be. There are plans underway to drill deeper, extending the data to 1.5 million years.*

A typical view of air bubbles trapped in ancient ice appears magnified here. Approximately 10% of ice core samples is trapped air. Not all of the air with its carbon dioxide is trapped in the bubbles themselves though. For example, we will mention that carbon dioxide can form hydrated species (see Chapter 10). Consequently, great care is taken to be sure that all gas is extracted from each zone of the sectioned ice core tube.

---

* D. Dahl-Jensen, "Drilling for the oldest ice," *Nat. Geosci.* 11, 702–706 (2018).

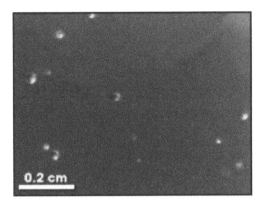

The comparison of the deuterium fractionation time series to that of carbon dioxide's is extremely good evidence in support of invoking the proxy: the very strong interrelationship between temperature and carbon dioxide levels (even putting aside the question of cause and effect). Furthermore, analysis suggests (and even a quick look at the graph reveals) a repetitive cycle time of approximately 100,000 years, plausibly associated with Earth's long-term orbital motion about the sun, as suggested by the Serbian physicist Milutin Milanković in the early 1900s. Average CO_2 levels bounced between about 180 ppm and 300 ppm while temperature *variations* are implied, from the deuterium proxy, to range between –9 to +3°C relative to today's global average. Minimum temperatures inferred from the above cycles match up with known and implied ice ages, especially the most recent one at 20,000 years near the left end of the figure and expanded in the next figure. Of course, there are fewer than 1 million years' worth of cycles shown in the figure, leaving a legitimate and interesting question as to just how regular the characteristics of the implied "Milanković" oscillations have been throughout the billions of years of Earth's existence.

Analysis of the composition of ice-bound air is arguably the most accurate and least problematic method of characterizing ancient CO_2 history.

Since the last ice age, the recent upswing emerges in the topmost ice layers. The correlation between atmospheric carbon dioxide content and temperature is consistent with warming over that time span. Modern measurements of atmospheric CO_2 levels have exceeded by another 100 ppm the typical cycle maxima shown in the analysis of bubbles shown here. Carbon dioxide has achieved that recent high at a very rapid pace as illustrated in the graph. (There is equally good data going back 45,000 years showing carbon dioxide levels staying between 180 and 220 ppm.)* Except for the most recent half century, the data is mostly from Antarctica.

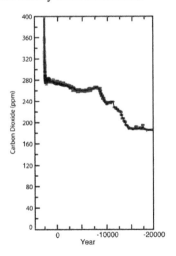

---

* J. W. B. Rae et al., "CO_2 storage and release in the deep Southern Ocean on millennial to centennial timescales," *Nature* 562, 569–573 (2018).

A clear increase in atmospheric carbon dioxide from its minimum 190 ppm to 260 ppm over the period from about 20,000 years ago to 10,000 years ago is associated with our emergence from the last ice age. That is, the significant increase in carbon dioxide over this time span is expected to reflect an accompanying increase in temperature for the region explored. The CO_2 level can be invoked as a proxy for temperature. Near the middle of this stretch, a break in the upward trend at around 12,000 years ago is a much-studied cool period referred to as the "Younger-Dryas." A number of different, independent proxy measurements have been made with extremely good refinement in terms of the age assigned to the data obtained. In the data graphed next, the topmost data trend over time shows the degree of grayness of an ocean bottom sediment core that had been extracted off the coast of Venezuela. Immediately below the Gray Scale plot is a collection of proxies exhumed from an ice core from thousands of kilometers away, in Greenland. The first of these (below grayness) represents accumulation of snow as a function of age. Below that is methane (CH_4) content in trapped air bubbles as a function of age, probably an indication of wetland productivity in which methane is biologically produced under proper conditions. Next is sodium ion content, most likely from sea salt blown in by winds and thus a proxy for wind transport. Last, calcium ions, also conveyed by wind transport, are interpreted as an indication of surface dust associated with dry (and windy) periods. This particular collection of results has been introduced to illustrate the idea of proxies and how matched behavior among assorted stand-ins can provide a sense of confidence for an interpretive idea: in this case the 1,300-year long cool "Younger-Dryas" is also referred to as the "Big Freeze."

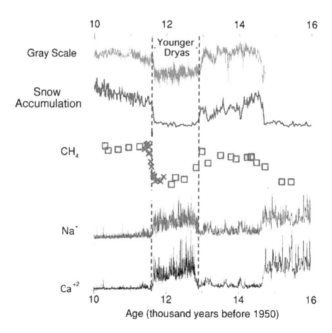

Also measured from the Greenland ice core whose analysis was presented earlier were some isotope fractionations in the ice (H_2O) itself, leading to a recreation of the temperature change over the period covered by the core. That analysis is pictured in this graph of temperature versus age. (Note, the time scales on the horizontal axes of the two figures do not match.)

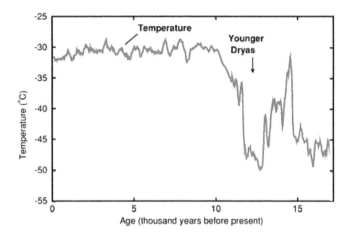

CAVE DROPPINGS

The fancy name here is *speleothems* for cave deposits, mostly "dripstones" such as *stalactites* hanging from cave ceilings and *stalagmites* reaching up from cave floors. Beautiful speleothems, very slowly grow from drippings of rainwater containing dissolved carbon dioxide and minerals. Rainwater seeps underground, perhaps over large deposits of aragonite or calcite (limestone) deposits of calcium carbonate. The limestone can dissolve if conditions are right, depending on how much carbon dioxide is in solution, on the rainwater acidity, on calcium ion content, on temperature, and other factors. As more and more limestone is carried away in solution, cavities appear and these can eventually, over thousands of years, grow into huge subterranean caves extending for mile after mile. With seepage of rainwater continuing, and again, under proper and not unusual circumstances, calcium carbonate will re-precipitate out of solution. Fluid initially confined in compact, moist soil then enters open chambers allowing evaporation of water, concentrating the minerals within the solution.

Among the dissolved substances are natural traces of uranium and thorium that have been leached out of surface soils. These are radioactive and can become trapped and mineralized in the calcium carbonate deposits serving, if probed and analyzed, to give accurate ages of the deposits as a function of the extent of very slow accumulation of the stalactites and stalagmites. The amount of uranium and thorium is not critical since it is the ratio of isotopes connected by radioactive decay that serves as the timing clock and those ratios can be measured by extremely sensitive techniques.

Layer after layer of calcium carbonate can be sampled and the isotopic fractionation of its elements can be quantified, serving as a proxy of chemical influences over the ages: an archive. As one exquisite example, the oxygen fractionation in $CaCO_3$ deposits from central China, about 700 km west of Shanghai in the Sangbao cave has been investigated. This is a region of China that, over the ages, has experienced sporadic severe monsoon seasons, including in the present. Some 80% of the annual precipitation (2 m) occurs during summer. How this is relevant is simply the fact that monsoons dump a lot of water containing the various forms of carbon dioxide whose isotopic composition represents, by proxy, atmospheric conditions. The graph here shows the relative change in enrichment of the heavy oxygen isotope ^{18}O in cave carbonates compared to a standard. The period shown traverses the end of the second ice age (a "termination" represented by the label T_{II} in the earlier figure of CO_2 content from glacial ice bubbles). The effect is extraordinarily dramatic and matches not only with the ice-core results, but also with other proxies: marine isotope fractionation studies and ice-rafted dropstone debris recovered from the North Atlantic. Note that the alleged signature for the monsoon imprint has been dated at 129,000 years ago with an uncertainty of only

100 years. Similar isotopic signatures have been observed for the ends of glacial terminations I, III, and IV as well, but with less precision in ages.

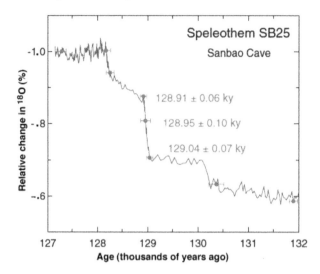

READING ABOUT THE ATMOSPHERE IN TEA LEAVES

For unraveling past situations, a technique that is complementary to the use of assessing changes in isotope abundances due to physical and chemical conditions – isotope fractionation – involves plant leaves and their "stomata." Stomata are near-microscopic mouth-like structures on leaves. Stomata can open and close, depending both on carbon dioxide levels and on the need to retain water. A magnified picture of a leaf surface showing some stomata is displayed here.

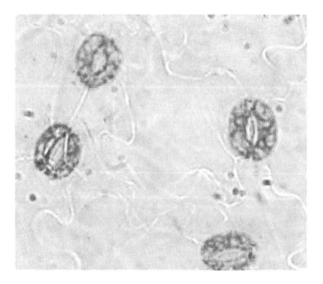

Certain species of plants, including especially the *gingko biloba* that is used in herbal tea, respond to carbon dioxide levels by varying the fraction of surface cells that are stomatal cells. This "stomatal index," a proxy for CO_2 in the atmosphere, has been found to be nearly independent of other external factors such as temperature, light, and nutrient supply. Ian Woodward of Sheffield, England, first conceived the idea in 1987 to measure stomatal changes as a proxy for carbon dioxide.

Hypothetically, changes can occur over a single year, thus providing exceptional time resolution if leaves can be found and dated. More commonly, though, long-term changes in carbon dioxide levels are reflected as changes in the average stomatal index of buried or fossil plants. Leaves from the past one-and-a-half centuries were used to quantify the relationship between stomata data and carbon dioxide levels. Whether or not other factors unique to this recent time period can affect interpretations remains arguable. Among those factors are the availability of water and variations in light intensity.

Dependence of the stomatal index on CO_2 levels has to be determined (calibrated) carefully under controlled laboratory conditions. Since plant behavior is different among even similar species, the optimum situation is to work with the same plant species over the time period being explored. A European study of birch dating back about 10,000 years suggests a CO_2 level rising from about 265 ppm to 327 ppm in less than 100 years, an increase of 23%. The earlier figure shows a lesser increase in Antarctica and a lower concentration as well.

As it happens, there are a number of plant species that are understood to have evolved only slightly over even hundreds of millions of years. The *gingko biloba*, a species believed to have been in existence for millions of years, is informally referred to as a "living fossil." Trees such as conifers and birch are a few more such examples. Their abundant, well-preserved remains retain impressions of stomata. Their stomatal indices can be determined readily. Using calibrations determined from modern but presumably very similar species under various CO_2 levels, the indices can be converted into apparent carbon dioxide atmospheric concentrations. A joint British–U.S. study of stomatal indices of *lycopsids** dating back to the Carboniferous glaciation period of around 300 Mya implies average carbon dioxide atmospheric concentrations of a bit more than 300 ppm. This value certainly seems to disagree with results with some carbon isotope studies mentioned earlier, being a factor of 2 lower. Trends, rather than precise absolute values, are quite reliable. The stomatal method has very good time resolution, perhaps a hundred years. But to reveal a trend, samples are needed that cover a long period of time continuously. This is frequently not the case for the distant past. One way to circumvent some of the ambiguities is to use multiple species, each independently calibrated.

SOILED RECORDS

Calcite ($CaCO_3$) forms in soils, especially where evaporation of moisture exceeds the precipitation rate as is the situation in arid or less-than-humid regions. With evaporation, the concentration of dissolved minerals increases, approaching levels where the solubility limit is exceeded (see Chapter 12). The calcium is provided by minerals in windblown dust from which dissolved ions are carried downward into the soil by rainwater. Eventually, the Ca^{2+} ions can combine with bicarbonate ions that have been generated from carbon dioxide that has diffused into the exposed soil layer. Calcite precipitates.

$$Ca^{2+}(aqueous) + 2\,HCO_3^-(aqueous) \rightarrow CaCO_3(solid) + CO_2 + H_2O$$

The technical label for this calcite is *pedogenic carbonate* where pedogenic refers to processes occurring in soil or forming soil. The soil carbonates provide an archive of CO_2 isotopic composition. That composition, in turn, is governed to a knowable extent by atmospheric carbon dioxide at the time of equilibration and to another extent by contribution from respiration of organic soil material by microbes, a contribution that can be estimated. What serves in the estimations and evaluations of pedogenic carbonates is the isotope fractionation of the carbon. That is, $^{13}C/^{12}C$ measurements in carbonate precipitates deposited in soil at different depths can be analyzed to disclose

---

* *Lycopsids* are presumed to be the oldest group of living vascular plants, that is, those having circulation vessels, whose extinct members played a major role in the extension of plant life onto land, eventually leading to the explosive growth of plant life that contributed to the coal beds being mined today. Among the *lycopsids* are some mosses and quillworts.

what the atmospheric pressure of CO_2 was at the time the soil layer was in equilibrium with the air. Of course, that assumes equilibrium was achieved.

The rate at which calcium carbonate forms is rather slow. Consequently, part of the attributes of the pedogenic carbonate proxy technique is that the time period for a particular sample will be spread over thousands of years.

Conventionally, carbon dioxide in the atmosphere equilibrates with calcium ions in water and can yield calcium carbonates, calcite, and aragonite in particular and especially in alkaline, saline lakes. But the sodium ion, Na^+, is much more abundant in the aquatic environments and almost everywhere than Ca^{2+}. From that, it seems obvious that sodium carbonates would form. But sodium carbonate (Na_2CO_3) is much, much more soluble in water than the calcium compound and does not precipitate under ordinary circumstances to form mineral deposits. If the water evaporates, however, these carbonate solids can form. Their identity depends on temperature, but also on the atmospheric pressure of carbon dioxide. Ordinarily, two forms predominate. *Natron* is $Na_2CO_3 \cdot 10H_2O$, a hydrate of sodium carbonate, and is dominant at low to moderate temperatures. At higher temperatures, *trona* ($NaHCO_3 \cdot Na_2CO_3 \cdot 2H_2O$) can prevail. Yet a third form, the rare *nahcolite* ($NaHCO_3$) can appear. Experimentally, nahcolite will not occur until atmospheric carbon dioxide is at least 550 ppm…unless NaCl salt precipitation occurs as well. In this latter circumstance, the carbon dioxide level would have to exceed 1,125 ppm. Consequently, deposits of nahcolite are proxies for high carbon dioxide levels. Deposits of nahcolite up to 300 m thick occur in a few regions and date to 45–55 Mya.*

ALKENONES, MARINE ALGAE COMPOUNDS

Phytoplankton – marine algae – are responsible for half of the photosynthesis that occurs in the world today, extracting carbon dioxide near the sea surface where sunlight powers chemical transformations that release oxygen into the atmosphere and initiate sequences of reactions building organic molecules. There are over 5,000 species of phytoplankton known. Among the very unusual and seemingly unique chemicals manufactured by phytoplankton are *alkenones*, two of which ($C_{37.3}$ and $C_{37.2}$) are shown here. Their production can be easily studied in laboratories today. These algae are found also in ancient marine sediments. Of course, some uncertainty is acknowledged because of unknown evolutionary changes in the algal biochemistry.

Of interest here is that these alkenones are very resistant to decomposition and decay over periods as long as 100 million years. Moreover, the amount of ^{13}C incorporated into the molecules has been shown experimentally to be very sensitive to the amount of carbon dioxide in the phytoplankton's living domain, the ocean surface. Because the sea surface carbon dioxide is in equilibrium with adjoining atmospheric carbon dioxide, the study of alkenones is a fingerprinting method for determining how much carbon dioxide was in the atmosphere at the time of the phytoplankton's life. Upon death, phytoplankton sink to the seabed, contributing to ever-deepening layers. Ocean drilling has recovered samples from several global sites and has assigned dates to the different depths. Yet, there are hurdles with the alkenone proxy. Measured quantities also depend on nutrition

---

* T. K. Lowenstein and R. V. Demicco, "Elevated Eocene atmospheric CO_2 and its subsequent decline," *Science* 313, 1928 (2006).

availability, so independent means are needed to show that "food" supply was either constant or its trend available through other measurements. Likewise with temperature effects on isotope fractionation in alkenones. Finally, the question also remains as to the extent of evolutionary changes on alkenone generation between current day calibrations and physiology of ancient phytoplankton.

READING ABOUT THE ATMOSPHERE IN AMBER

Pliny the Elder, a Roman naturalist, author, philosopher, and naval officer living in the first century, was one of the first to recognize that amber containing insects was a fossil resin of pine tree saps. Fossilized insects were among the many lifetime observations he recorded in his 37 volume *Naturalis Historia* written between the years 77 and 79.

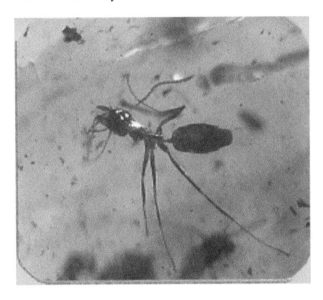

As we have seen, ice cores provide physical and chemical evidence of past times going back for fewer than a million years. Bubbles of air trapped in amber – those oozing resins from plants that occur throughout the world – conceivably allow a glimpse as far back as the Carboniferous Period some 300 Mya.

The basic idea here is quite elegant. Tree resin that has hardened into a glass-like substance seals "airtight." It is thought that plants use the resins to close off wounds or to occlude invading insects or microbes. Any trapped gases remain (hypothetically) unaffected by the passage of eons since the resins are quite resistant to degradation over time. Samples of amber gathered from a variety of places are crushed in a lab environment to release the trapped gases. Measurements take advantage of analytical instruments sensitive enough to detect fewer than ten parts per billion of gas from bubbles as small as a ten-thousandth of a centimeter in size.

Results published in 1988 interpreting amber contents suggested that the oxygen content of the atmosphere has changed appreciably over millions of years. But some obvious concerns with the use of amber gases is that microorganisms present in the substances could metabolize oxygen for a while. If that were so, however, respiration metabolism would produce carbon dioxide from consumed oxygen on a one-to-one basis. In such a situation the sum of oxygen plus carbon dioxide contents would remain constant as a percent of the total gas released. In modern ambers, the ratio of oxygen plus carbon dioxide to total gas was found to be 21% in perfect agreement with environmental values (oxygen is 21% and carbon dioxide is less than 0.04%). That study, too, at least partially supports the thought that there is no leakage into or out of bubbles, especially if samples of the same ancient age show nearly identical oxygen plus carbon dioxide content.

Nevertheless, use of amber as a means to identify the composition of ancient air has met with some criticism. For example, any gases produced by tree metabolic processes and accumulating in the resin would have nothing to do with trapped atmospheric air yet could distort the composition in significant ways. Furthermore, any differences in the solubility of various atmospheric gases in the resin/amber or of chemical reactions even upon pulverizing the samples would alter the gas ratios. Such problems are further confounded by the fact that different ambers have different compositions themselves, varying with geographical origin and also with age. Finally, some, but only some, studies of how long gases can reside in amber have given times of only several years. If so, the exchange of bubble contents with the external environment perplexes interpretation.

Amber analysis, supported by chemical composition analyses, pollen identification, bacteria, algae, and fungi identification, is an elegant idea for probing the remains of ancient atmospheres but has yet a bit of convincing to do.

In the year 79, Pliny the Elder died on a ship trying to rescue family friends from volcanic eruptions in nearby Pompeii. Reportedly he succumbed to toxic fumes, probably carbon dioxide.

The past is a foreign country; they do things differently there.

LESLIE HARTLEY
British author

9 Fire

Double, double toil and trouble; Fire burn, and cauldron bubble.

WILLIAM SHAKESPEARE
(*Macbeth*)

FORESTS

Almost 99% of forested land is in a "carbon dioxide uptake" mode. The life expectancy of trees – trees are where most of the carbon dioxide is terrestrially used – ranges from 50 years up to a few centuries. But uptake is slow. Especially in comparison to release. Carbon dioxide can also be *produced* by trees rather than merely being consumed (Chapter 15). Emissions by forests include processes such as respiration, tree death and decay, timber harvesting, and pest outbreaks. Fires are the most dramatic release process. Of course, fires, although actually rare in a sense, can liberate 200 years' worth of sequestered carbon dioxide back into the atmosphere in a flash. Forest fires can generate temperatures as high as 2,000°F (1,100°C) and consume upward of 35 tons of fuel per acre each hour comparable to 10% of the daily average consumption of fuel oil in the United States. Estimates of emissions from wildfires are problematic owing to variations in the type or the chemical composition of combusted material, structure, condition, and meteorological conditions during a wildfire.

For reference, anthropogenic (human caused) carbon dioxide sources release 34 gigatons per year. More than half is from fossil fuel combustion: about 20 gigatons per year. Regionally, fuel emissions have been varying significantly, increasing in some regions – China and the Middle East – and decreasing in the Americas as shown herein a graph based on data from the International Energy Agency 2011 publication "CO_2 Emissions from Fuel Combustion: Highlights."

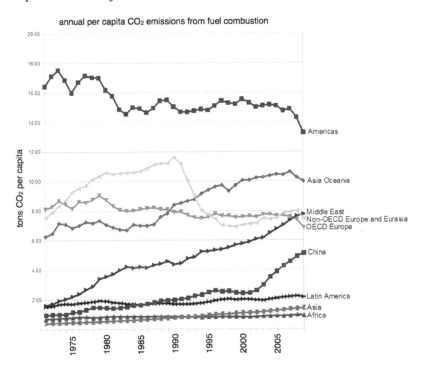

annual per capita CO_2 emissions from fuel combustion

Carbon dioxide is also generated in large amounts from the manufacture of cement, but this represents only 1% of the man-made total. Most of the remaining production is from global biomass, the major contribution is forest fires. These sum to perhaps 14 gigatons of carbon dioxide per year or more than one-third of the total or a nearly 2 ppm additional atmospheric pressure. In arriving at these estimates, it is frequently necessary to quantify peat (the compact deposit of partially decomposed vegetation) and soil as additional sources of CO_2 released by fires. These latter estimates are somewhat problematic. Estimates indicate peat deposits can range up to 20 m in thickness.

For 1995 worldwide, the estimate is that greater than 1,900,000 square miles burned. This is equivalent to two-thirds the area of Australia, for example, and amounts, on the average, to *all* global land (exclusive of Antarctica) burning each generation. In wildland areas, an estimate from not too many years ago was that four-fifths of fires were initiated by lightning. (There is only a one-in-a-hundred probability that any given lightning strike will ignite a fire.) Today, it is recognized that the vast majority of fires in forests, savannas, and grasslands are intentionally set. Reasons include timber harvesting, land conversion, wood for cooking and heating, and slash-and-burn agriculture. The table shows a 1991 estimate of sources of biomass burning and the amount of carbon dioxide released in a year. In a few cases, the estimated total area consumed by fires is indicated as well.

Source	CO_2 Released (Billion Tons/yr)*	Area (1995 Estimate) in Sq. Mi.
Savannas	6.09	1.9 million
Agricultural waste	3.30	
Tropical forests	2.09	75,000–150,000
Fuel wood	2.35	
Temperate and boreal forests	0.48	40,000–60,000
Charcoal	0.11	
World Total	14.40	

* M. O. Andreae, "In Global Biomass Burning: Atmospheric, Climactic, and Biospheric Implications," J. S. Levine, ed., MIT Press, Cambridge (1991) derived from a NASA report (vide infra).

Variation of contribution domains over more recent years is displayed here in the graphic.*

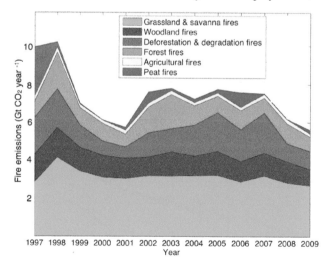

* G.R. van der Werf et al., "Global fire emissions and the contribution of deforestation, savanna, forest, agricultural, and peat fires (1997–2009)," *Atmos. Chem. Phys.* 10, 11707 (2010).

Major fire types unsurprisingly vary with geographical location. The distribution of dominant classifications is shown globally in the map presented here.*

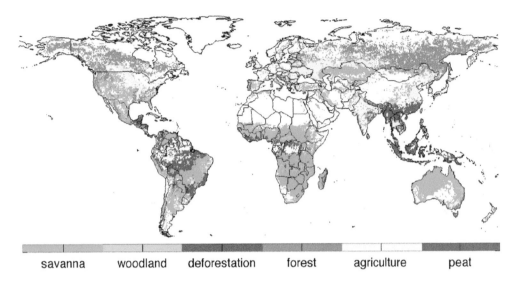

savanna   woodland   deforestation   forest   agriculture   peat

In many areas, production of carbon dioxide from fires is nearly balanced by uptake over subsequent years of regrowth. Furthermore, combustion of carbon-containing materials does not necessarily convert all of the carbon into carbon dioxide. Charcoal and carbon monoxide (CO) are common products as well, typically amounting to 25–30% of the total, and depend on multiple factors including fuel classification shown here as kg product per kg dry matter combusted.

	Deforestation	Savanna and Grassland	Woodland	Extratropical Forest	Agricultural Waste Burning	Peat Fires
Carbon	0.49	0.48	48	0.48	0.44	0.56
CO_2	1.63	1.65	1.64	1.57	1.45	1.70
Carbon monoxide	0.10	0.06	0.08	0.11	0.09	0.21

BOGS

In recent years, there have been major forest fires in Indonesia and Brazil. In the former, in just a few months in 1997, it is estimated that at least 18,000 square miles of forest on the islands Kalimantan (Indonesian Borneo) and Sumatra burned, generating greater than 700 million tons of carbon dioxide, perhaps 3.5 billion tons if all islands are included, an amount equal to that of Europe's anthropogenic formation for a year. Most of the sources, about 90%, were combustion of peat, thousands of years accumulation of the partially decomposed remains of marshland vegetation. In Borneo, the peat bogs were uncharacteristically dry because of climate fluctuations and also because of government wetland drainage activities in the region. Estimates are that this colossal fire released perhaps as much as 40% of the global production of carbon dioxide from fossil fuels in a year and is thus extremely significant. By comparison, the Kuwaiti oil fires (shown in the photograph here) after the 1991 Gulf War released between 100 and 500 million tons of the gas.

* See previous footnote.

The satellite projection on this map shows the smoke and soot dispersal over the island of Borneo (center of map). When inquiring about the extent of the burnt area from the literature, values that differed by more than a factor of 2 from each other were not unusual findings. There is much uncertainty in all these data because a complete and precise assessment is apparently very difficult to acquire. The land area addressed here, amounting to almost one-third of the area of Borneo, is believed to be reliable since it originates from broad satellite imagery although smoldering peat fires can be difficult to detect since they can occur at lower temperatures, high moisture content, and low oxygen concentrations. In 2002, the fires started up again and have been erupting annually, squelched only by seasonal rains starting each October.

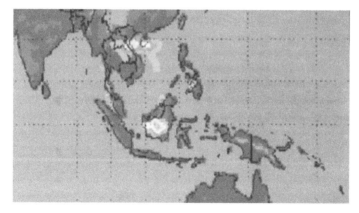

Here is a more recent satellite image of the residual haze.

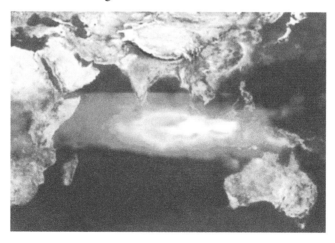

As far as Brazil goes, the estimate is that in 1998, following an unusually low humidity period, 100–200 megatons of carbon dioxide were generated by burning forests. The years 1997–1998 correlate with one of the largest increases in carbon dioxide emissions.* Presented here is a satellite-generated image of the fire's location in Brazil. Presently, nearly 60% of the Amazon region is too humid to support deforestation fires though.

August 27

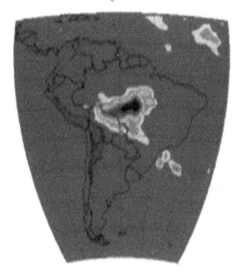

The largest tonnage of carbon dioxide released by combustion comes from savannas, three times that from forests. Since about two-thirds of savannas are in Africa, that continent gets referred to as the "burn center" of the planet. A satellite image of fire locations in Africa is shown here.

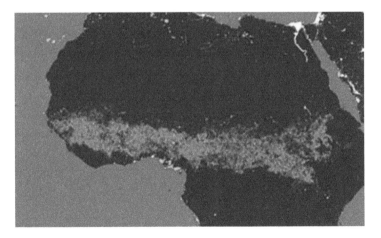

Not all fires are equal though. The area burned does not necessarily translate into fuel consumed and carbon dioxide released. In old forests, fuel has accumulated and is dense. In African savannas, huge areas can burn, but the grasses and bushes are sparse suppliers of fuel. True, savanna totals are triple those of tropical forests but burnt areas are actually 20 times more extensive than those of forest burns. Additionally, slow burning fires are particularly difficult to quantify, especially in soils with deeply embedded organic material.

---

* D. Schimel, "Carbon cycle: the wildfire factor," *Nature* 420, 29–30 (2002).

One of the largest fires ever measured was in May 1987 in northeast China and into bordering Siberia. In under a month, more than 5,000 square miles were burned. Fires extended across the border into areas of Russia as well. Satellite images suggest that eventually a total of over 50,000 square miles were consumed here during 1987. A half billion tons of carbon dioxide were released. This could be compared to the 1980 estimates of 30,000 square miles of boreal forest burning annually (with great fluctuations in what that represents). An even earlier estimate of 6,000 square miles total per year has been made. Prehistoric records of fires, even indirect proxies, not surprisingly, are absent. Whether the large value of 50,000 square miles is an aberration or a trend is simply not known. Siberian "heat waves" during the summer of 2003 spawned wildfires that incinerated about 85,000 square miles, releasing about 1 gigaton of carbon dioxide or 6% of the average global annual total for fires. Most terrestrial fires are in the northern hemisphere (which includes the African savannas). Siberia is about 10% of that terrestrial domain.

By comparison, in the United States, major wildfires over the past decade in Florida, California, and New Mexico consumed ~100,000 acres or ~150 square miles per conflagration. Annual total conflagrations range 4–10 million acres and are known to vary from year to year. One of the worst years for fires in the United States has been 2018 when cumulatively over 13,000 square miles burned in 10 months. The National Fire Information Center reports the 10 year average for 2008–2017 was 58,000 fires and 6.3 million acres (10,000 square miles) consumed.

The map of the world shown here divides the global landmasses into several geographical regions. "EQAS" stands for equatorial Asia and includes the island of Borneo whose 1997 bog fires were discussed earlier. The subsequent display panels show, for EQAS in the lower left, sharp spikes in carbon dioxide production by fires in the year 1997 as well as 2002 and 2006 to lesser extents. This occasional peak CO_2 production can be contrasted to that for "NHAF," northern hemisphere Africa, where seasonal peaks regularly produce 200 million tons of carbon dioxide in December. In southern hemisphere Africa, below the equator where the seasons are reversed, peak production by fires occurs midyear with slightly lower values. (Be sure to note the changing vertical scales for each panel.) The broadly defined Middle East region "MIDE" has the lowest peak monthly product of carbon dioxide from fires.

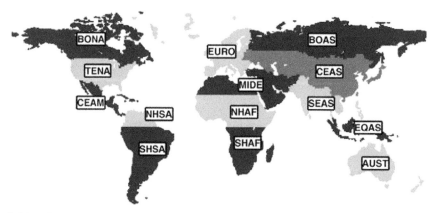

BONA	Boreal North America	NHAF	Northern Hemisphere Africa
TENA	Temperate North America	SHAF	Southern Hemisphere Africa
CEAM	Central America	BOAS	Boreal Asia
NHSA	Northern Hemisphere South America	CEAS	Central Asia
SHSA	Southern Hemisphere South America	SEAS	Southeast Asia
EURO	Europe	EQAS	Equatorial Asia
MIDE	Middle East	AUST	Australia and New Zealand

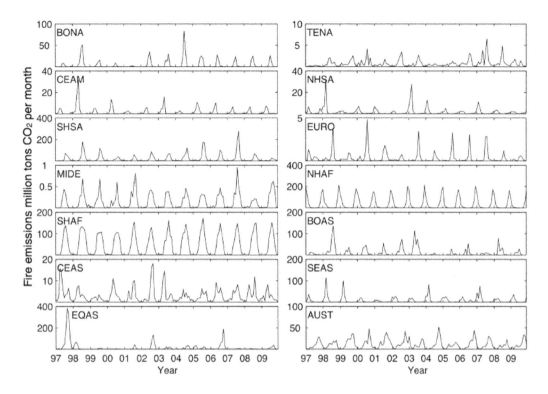

Perhaps a clearer way of viewing the production of carbon dioxide from fires is to globally map its local production, derived from satellite measurements, averaged over the 13-year period ending in 2009, smoothing out the fluctuations from year to year. This is illustrated by the graphs here. The paucity of fires associated with desert or desert-like areas in Africa, the Arabian Peninsula, and the Gobi, for example, is expected.

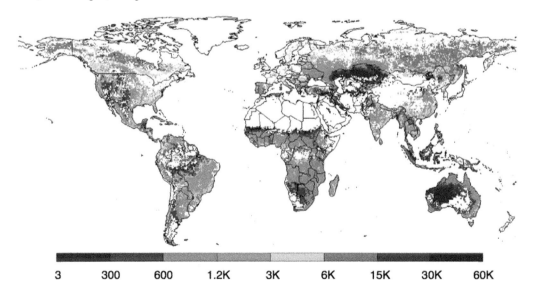

MEGAFIRES

Runaway plant growth during the Carboniferous Period 300 Mya (Chapter 7) voraciously consumed carbon dioxide, generating oxygen via photosynthesis (Chapter 14). If combustible carbon is buried (sequestered) permanently, oxygen levels eventually rise. Vast coal deposits throughout the world support this picture for the Carboniferous Period. Taking carbon out of carbon dioxide leaves oxygen. Some evidence and modeling suggest that during the Carboniferous Period oxygen levels were, indeed, significantly higher than now (see the figure in the previous chapter): perhaps as high as 35% *versus* 21% today. Burning limitations associated with 35% oxygen in air were supported by early experiments that turn out to have been performed using thin paper to model a combustible substance. On the other hand, it is known that wet organic material can burn to some extent in air containing >25% oxygen. If so, lightning strikes could ignite fires even in rainforests, devouring oxygen and returning carbon dioxide to the atmosphere as carbon dioxide. Wildfires, if uncontrolled, could have decimated vegetation. This conceivably puts a natural brake on rising oxygen levels and imposes an upper limit on atmospheric oxygen content. But to indicate how intricate it is to invoke these balances, the "wild fire" would need to return basically all the organic carbon to the atmosphere. But in dense forests following such fires, there invariably remain immense quantities of charcoal, a familiar form of nearly pure carbon that is essentially impervious to attack (metabolism, respiration) by microscopic organisms. Very recent investigations of burning wet organic matter in high oxygen environments showed that fire likelihood is not necessarily that influential in restricting atmospheric oxygen concentrations to certain maximum levels. Many plant species with high wood (lignin) content do not burn well, leaving behind unburned char fractions of 40% or even more. Furthermore, the high water content of natural vegetation is not conducive to supporting *uncontrolled* combustion in environments having 35% oxygen. Given this possibility, fires can engender further sequestering of carbon, enabling the production of yet more oxygen, a contrast to intuitive predictions. Naturally, as oxygen rises yet further, the likelihood of fire increases accordingly.

GLOBAL CONFLAGRATION

The "mother of all fires" has been a notorious subject of interest, investigation, and controversy for nearly two decades now. The evidence is focused around an asteroid impact 66 Mya. The asteroid struck the Yucatan in Mexico with a speed some 40 times the speed of sound. The solid mass had a diameter estimated to have been such that when its leading surface first touched the ground, its opposite side was at the height where a typical commercial jetliner would cruise nowadays. Namely, we are talking about an object that was 10–15 km in diameter. An explosion equivalent to the detonation of 100 trillion tons of TNT ensued, wreaking havoc everywhere. Much of the data for this impact comes from deposited layers dispersed in dozens of sites globally and, from dating – consistently – by a variety of independent techniques, to 66 Mya. Even 7,000 km away from ground zero, layers of ejected material deposited in accumulations up to 2–3 mm thick are found. *T. Rex and the Crater of Doom* by W. Alvarez and C. Zimmer gives an excellent exposition on this subject.

In 1985, a young scientist named Wendy Wolbach discovered a single, very thin layer of soot in most of the deposits associated with the impact. The chemical composition of the enriched soot layer was similar to that found following forest fires. The total amount of consumed carbon implied by this ubiquitous layer added up to an astonishing 10% of today's total biomass: trees, and so on. Layers immediately above and below had four times less soot, if even that much. High-resolution microscope views showed structure characteristic of the kind of carbon (soot and carbon black) typically generated by flames. Slightly larger aggregates resembled charcoal associated with forest fires. As a result of these forensic studies, a second act was added to the dramatic event of the destructive Yucatan collision. Massive amounts of extraordinarily hot debris were ejected into the atmosphere, spreading out over vast areas, heating the air, drying vegetation, and settling down onto limitless supplies of kindling worldwide. Organic matter (peat) in topsoil and other fossil fuels such as oil shale near the surface would nourish the conflagration. Vast wildfires would erupt almost everywhere simultaneously. Furthermore, the immense heat created at the vast impact site, a site rich in carbonates such as limestone, would have added more carbon dioxide from thermal decomposition

$$CaCO_3 + heat \rightarrow CaO + CO_2$$

Estimates are that 10 trillion tons of carbon dioxide, in addition to other gases, were generated, pumping up the atmospheric levels to many times their typical values. Soot and smoke darkened the skies. By how much is uncertain. But recently,* more detailed evidence, in which specific organic residues (polyaromatic hydrocarbons) were identified, strongly supports a yet more intricate explanation. The blaze was fueled by the combustion of hydrocarbons and organic matter known to have had a significant presence in the rock at the impact site. And residues of hydrocarbon incineration serve as proxies for the event.

Between the greenhouse effect of the added CO_2 and the lack of sunlight, major extinctions of life forms were inevitable. This hell-on-Earth ended the reign of the dinosaurs. The accumulated evidence is extremely compelling.

Ocean sediments settling above the alleged impact layers switch sharply from carbonate rich to carbonate poor. Carbonate sediments are proxies for marine species. A very reasonable interpretation is that production of biologically generated carbonate basically ceased as a result of the asteroid cataclysm. Among other effects, increasing atmospheric carbon dioxide increases the amount of CO_2 in the ocean resulting in a more acidic medium with decreased concentrations of the carbonate ion, CO_3^{2-} (see Chapter 10). Abundant skeletal fossils in the sediments are also decimated at impact time. Taxonomy, the study of species, reveals that new species (nonexistent in older deposits) appear in higher, more recent sediment layers than those at collision time. Analyzing the new species for preference of cold *versus* warm climate also signifies a cooling following the impact due to the production of thick, dense, long-lasting cloud cover.

From where did the eventual rebirth spring? Earth was not torched everywhere. Some high latitude regions, near the poles, would have escaped the wildfires and remained relatively cool enough for some hardy species to survive. Slowly – just how slowly is conjecture – plants and animals began to expand their territories again.

And here we are.

COAL FIRES

Maybe 1,000 miles northwest of the Nyos and Monoun volcanoes of Cameroon, Africa, in the northern part of Mali, are some areas of ground that are exceedingly warm to the touch. They've been known to Western scientists for only just over 100 years and were logically assumed to be the result of underlying volcanic activity. Not all scientifically logical deductions necessarily pan out under detailed investigation though. Norwegian investigators dug an exploratory trench into the hottest seared ground and found an enflamed layer of peat with a temperature measuring 830°C buried at only a meter's depth. Digging deeper though, the peat temperature dropped to 40°C, demolishing the hypothesis that subterranean volcanic activity was the cause. Compounding the finding was recognition that the heat, common during dry periods, would fade away during wet seasons. That is not a characteristic of volcanic activity. The understanding that emerged is the following. Subterranean peat fire is caused naturally by heat from bacterial respiration, akin to an infection or to what happens in compost piles. Microbial growth is facilitated by oxygen permeating through the peat, peat that becomes porous during arid conditions (when pores, previously water-filled, open up).

---

* Belcher et al., "Geochemical evidence for combustion of hydrocarbons during the K-T impact event," *Proc. Natl. Acad. Sci.* 106, 4112 (2009).

Yet more impressive than peat fires are coal fires, which can ignite at temperatures as low as 80°C. Many coal seams ignite spontaneously or directly from lightning strikes or indirectly from forest fires. Others are human caused. Coal seams tens of meters underground are known to have been burning for decades or centuries. Marco Polo's documents allude to "burning mountains along the Silk Road." Baked shale above coal seams has been "dated" by geologists to millions of years ago.* Despite the fact that many current underground coal fires have human causes too, it is clear this is mainly a natural phenomenon. Major sites are in China, India, and Australia but coal is found everywhere and so coal fires are global. The Australian combustion source, in New South Wales, has been estimated to be between five and six millennia old, with temperatures reaching 1,700°C underground. In China, evidence suggests ages of more than a million years. Coal production in China, amounting to a billion tons a year, meets three-quarters of that country's energy demand. Yet the fires consume about 1–2% of that annually, releasing 50–70 million tons of CO_2 into the environment. The broad expanse of the fires is seen in the map here. They extend into Russia and Mongolia as well. Current technology cannot feasibly reverse such extensive breeding situations. Even though much funding has gone into attempts to extinguish the fires, only a few attempts have been successful while other fires grow in dimension.[†]

Based on Kijk (1995, no. 8, p. 29)

Change is scientific, progress is ethical; change is indubitable, whereas progress is a matter of controversy.

BERTRAND RUSSELL

---

* C. Kuenzer and G. B. Stracher, "Geomorphology of coal seam fires," *Geomorphology* 138, 209–222 (2012).
† C. Kuenzer et al., "Coal fires revisited: The Wuda coal field in the aftermath of extensive coal fire research and accelerating extinguishing activities," *Intl. J. Coal. Geol.* 102, 75 (2012).

10 Carbon Dioxide and Water

In consequence of burning coal "spiritus sylvestris" comes into being. This spiritus, which was formerly unknown and cannot be kept in vessels, and cannot be converted into visible form, I call by the new name "gas."

JOHANN BAPTISTA VAN HELMONT
(Ortus Medicinae, 1648)

A quintessential feature of carbon dioxide is that it is the de-hydrated form of an acid, *carbonic acid* (H_2CO_3). Carbon dioxide could also be called the anhydrous form of carbonic acid. If one molecule of H_2O is removed from one molecule of carbonic acid, CO_2 remains. But when quantities of carbon dioxide are present in water, not only does the acid form, but also produced are the ions bicarbonate,* HCO_3^-, and carbonate, CO_3^{2-}. Relationships among these three species and their natural behavior is a major topic for exploration.

WATER

The very fact that carbon dioxide dissolves in water frustrated early attempts to study the gas, a gas known since the time of Joseph Black a quarter of a millennium ago as "fixed air." British scientist Joseph Priestley's first chemistry publication was on carbonating water. For this study, he was recognized with the Copley Medal in 1772, awarded by the Royal Society of London.

The Copley Medal was established in 1709 and among the previous winners were Steven Hales (1739) and Benjamin Franklin (1753). Hales was involved early on in the studies of carbon dioxide using his own clever device for collecting gases over water. But carbon dioxide is fairly soluble in water. Its detailed behavior as a dissolved substance is also somewhat complicated, partially due to its character as an acid: the ability to release H^+ into solution. Therefore, we are obliged to look briefly at acid behavior in water.

pH

It is fairly common knowledge that the measure of acidity is expressed† in terms of "pH." Acidity is an indication of how much hydrogen ion, H^+, is present in solution. At first, it might seem common sense to talk about how many hydrogens were present in a suitable volume such as a gallon

---

* Bicarbonate is formally called "hydrogen carbonate."
† pH from German *potenz* (power) + *H* (symbol for hydrogen).

or a liter. But that number turns out to be so incredibly large that a separate unit is used to facilitate communication. As we use *ton* to mean 2,000 *pounds* (or *metric tonne* for 1,000 *kg*, which is 2,205 pounds), the *mole* is a (really, really huge) number, a counting unit that is carefully defined as the number of ^{12}C atoms in exactly 12 g of ^{12}C. That number in 2018 has been defined to be 602,214,076,000,000,000,000,000 (or $6.02214076 \times 10^{23}$). The number of atoms, ions, or molecules in solution is then conveniently expressed as *moles per liter*. (A liter is about 1 quart in volume: 1.06 quarts, more precisely.) H^+ concentration can range over many powers of ten, from greater than 1 mole per liter in extremely acidic solutions to less than a hundredth of a millionth of a millionth moles per liter in extremely basic (alkaline) solutions. The pH scale quantifies the H^+ concentration by expressing it in powers of ten, but with the sign reversed. Thus 1 mole per liter is identical to 10^0 moles per liter;* 0.1 moles per liter is 10^{-1} moles per liter; and so on. The powers of 10 in these two examples are 0 and −1. With the sign reversal that is part of the definition of pH, these concentrations become pH values of 0.0 and 1.0. Values of pH like these correspond to very acidic solutions. The higher the pH, the less acidic the solution.

Pure water (H_2O) is ever so slightly dissociated. That is, some water molecules spontaneously break apart into hydrogen ions (H^+) and hydroxide ions (OH^-). Chemical symbols represent this rupturing and reassembly action with opposing arrows,

$$H_2O \rightleftarrows H^+ \text{ (aqueous)} + OH^- \text{ (aqueous)}$$

one implying the dissociation (to the right) and the other the re-association to the left, both of which are occurring prolifically in a liter of water. The average condition is such that the hydrogen ion concentration is 0.0000001 (or 10^{-7}) moles per liter, corresponding to a pH of 7. This concentration amounts to about two dissociated molecules at any one time for every billion intact water molecules. The ionic fragments are surrounded by whole water molecules, hence the more meaningful labels "H^+ (aqueous)" and "OH^- (aqueous)."

Expression of concentrations in terms of powers of ten is not restricted to whole numbers. For example, the pH of blood is 7.4 corresponding (by definition) to a hydrogen ion concentration of $10^{-7.4}$ (which happens to be 0.00000004 moles H^+ per liter or 40% that of pure water). On the other hand, stomach acid has a pH of about 1 meaning its hydrogen ion concentration is more than a million times greater than that of blood.

BUFFERS, ACIDITY, AND ALKALINITY

The pH of ocean water near the surface is almost always found to be between 8.1 and 8.3, less acid than pure, neutral water. That is, ocean water is slightly basic, or *alkaline* under these conditions. At this pH, the most prevalent of the various aqueous carbon dioxide species is bicarbonate, HCO_3^-, as will be discussed soon. Furthermore, if acid (or base) is added to a sample of this ocean water, the pH hardly changes. These two simple observations are quite important to understand. They involve a phenomenon called *buffering*.

To explore buffering, suppose we start with a liter (~ one quart) of pure water having a pH = 7.0. Let us consider adding just a milliliter of a strong acid solution such as hydrochloric acid (HCl) in water (the milliliter has its own pH = 1.0). Despite mixing so that the solution dilutes the acid a thousand-fold (one milliliter into a thousand milliliters), the pH would drop all the way from 7.0 (for the initially pure water) down to 4.0. In this example, the concentration of the strong acid was 0.1 moles per liter† and we used one-thousandth of a liter. For that milliliter (mL), we are adding then only 0.0001 moles of hydrochloric acid to produce the sharp pH change.

---

* Recall that any number raised to the zeroth power like $10^0 = 1$.
† This is a simplification for demonstration purposes.

Suppose we had a weak acid, which we'll abbreviate as HA, and its sodium salt, NaA, each of which was soluble in water. We prepare a mixture that contains 0.5 moles of each, the acid and the salt, in a liter of water. The salt, by its nature, would be ionized in solution

$$NaA \rightleftarrows Na^+ \text{ (aqueous)} + A^- \text{ (aqueous)}$$

and provide 0.5 moles of A^- (and the complementary 0.5 moles of Na^+ which, as essentially an unaffected bystander, we need not comment on further). On the other hand, HA being a weak acid means it doesn't provide much H^+ when put into solution. We could say that the weak acid remains almost entirely intact as HA in the water. But the salt's A^- can combine with some H^+ from acid and can also extract one H^+ from one water molecule since undissociated HA is the favored arrangement for this weak acid. In this latter case,

$$A^- \text{ (aqueous)} + H_2O \rightleftarrows HA + OH^- \text{ (aqueous)}$$

releasing base, OH^-. As it turns out, the resulting pH of this solution would be determined by how effectively HA behaves as an acid and by the ratio of the amount of A^- to HA placed in the liter of water. In our example,* that ratio is 1.0. If we chose the weak acid properly, the pH could be 7.0. (It is not really relevant, but the weak acid *hypochlorous acid*, HOCl, comes pretty close to satisfying this setup.) If to this pH 7 solution we now add, as in the earlier illustration, our 0.0001 moles of strong acid (as 1 mL of pH 1 HCl) thus disturbing the equilibrium situation, the H^+ from that strong acid would combine with plentiful A^- on a one-to-one basis forming HA and restoring equilibrium. Again, this is because HA as a weak acid prefers not to exist as H^+ plus A^- but rather as the bound combination. That means the 0.5 moles of A^- originally present when confronted with 0.0001 moles of H^+ are reduced to 0.4999 moles, and the HA that started as 0.5 moles is increased to 0.5001 moles. However, the ratio A^-/HA only changes to 0.4999/0.5001 = 0.9996 from 1.0000. The pH of the solution is hardly affected since it is determined by the noted ratio. The mixture of weak acid plus its salt served as a *buffer*, dampening the influence of the acid that had been so effective when added to pure water as previously demonstrated. If we added a strong antacid – a base – instead of a strong acid, the buffer similarly would keep the pH from rising more than a slight amount.

The ocean is buffered to an extent at around pH = 8.2. But what do we mean when we say it is buffered "to an extent"? In the earlier example of preparing a buffer mixture, if we instead used much less than the 0.5 moles each of HA and A^-, though keeping the ratio the same, the buffer would not have had as great a capacity in maintaining the pH as before. That's because the ratio A^-/HA would be more sensitive to changes. (What would happen if we had used 0.0005 moles each of A^- and HA instead of 0.5 moles each? After adding the 1 mL of strong acid, the A^-/HA ratio, [0.0005–0.0001]/[0.0005+0.0001], would drop to 0.67 from 1.00 compared to the previous 0.9996 *versus* 1 and the pH would adjust to 6.8 from 7.0.)

(Blood that circulates through most air-breathing living systems is also a buffer. The "HA" in this case is mostly carbonic acid and the "A^-" is the bicarbonate ion. The pH of blood, human blood at least, is around 7.4.)

The buffer capacity is referred to in some fields of study as the "alkalinity" of a complex solution. Although there are several different, yet individually precise, definitions of alkalinity in the technical literature, the alkalinity is essentially a measure of all those species in solution that could combine with H^+ to maintain a fixed pH thus acting to enhance buffer capacity. Salts, like those containing the hypothetical A^- mentioned earlier, can combine with H^+. Other salts might contain a "B^{2-}" which could combine with *two* H^+ to give H_2B. B^{2-} would contribute doubly to the alkalinity. Natural water systems' buffering ability is almost always described in terms of its alkalinity.

---

* 0.5 parts A^- and 0.5 parts HA gives the ratio 0.5/0.5 = 1.

ENTER CO_2

The natural buffering action of ocean water is very complex. Among the most influential buffering species is bicarbonate. The bicarbonate ion can pick up H^+ to form H_2CO_3 counteracting a potential drop in pH. Alternatively, loss of acidity, that is, gain in pH, is buffered by bicarbonate's other capability, its ability to give up an H^+ forming the carbonate ion, CO_3^{2-}. These two bimodal, opposing talents are represented, respectively, as

$$HCO_3^- \text{ (aqueous)} + H^+ \text{ (aqueous)} \rightarrow H_2CO_3 \text{ (aqueous)}$$

and

$$HCO_3^- \text{ (aqueous)} \rightarrow H^+ \text{(aqueous)} + CO_3^{2-} \text{ (aqueous)}$$

The complexity of water buffering is also affected when the ocean is in contact with sediment carbonates, calcium carbonate in particular. This will be discussed in Chapter 12, but for now, we'll just note the relevant buffering chemistry as

$$CaCO_3 + H^+ \text{ (aqueous)} \rightarrow Ca^{2+} \text{ (aqueous)} + HCO_3^- \text{ (aqueous)}$$

$$(H^+ \text{removal, resisting acidity gain})$$

$$Ca^{2+} \text{ (aqueous)} + HCO_3^- \text{ (aqueous)} \rightarrow CaCO_3 + H^+ \text{ (aqueous)}$$

$$(H^+ \text{release, aiding acidity gain})$$

Arguments based on proxies are quite convincing that seawater pH has been around 8 for eons. Any significant long-term "attempts" at acidification, propelled by CO_2 or even HCl from volcanoes, would (eventually) be significantly counteracted by dissolving copiously available $CaCO_3$ sediments (which have been around for hundreds of millions of years). Equilibrium takes time, though short-term changes – local or global – are certainly feasible.

CO_2 SOLUBILITY

The solubility of carbon dioxide gas in water at 25°C is 0.76 L of the pure gas per liter of water if the CO_2 gas pressure on the water is maintained at one atmosphere. (The gas volume 0.76 L corresponds to 0.031 moles, or 1.37 g, or 19,000 billion billion molecules of CO_2.) Keep in mind though, that adding 0.76 L of gas to 1.00 L of water does not give 1.76 L of solution, but ends up very close to the original 1 L volume, mostly because the number of molecules in the liquid has increased by only 0.05%. One atmosphere of carbon dioxide is not the norm. The pH of such a solution would be 4.0, about that of orange juice. If instead you added those 1.37 g to one liter of water and there was no carbon dioxide in contact with the water, some of the dissolved gas would leave the solution, reducing the dissolved CO_2 until equilibrium with the surrounding gas was established.

Although carbon dioxide dissolves relatively easily in water, only about 0.25% actually combines with H_2O to form the molecule H_2CO_3, the weak (and, having potentially two H^+ to contribute, *diprotic*) acid called carbonic acid. Not only that, but the change to carbonic acid is not instantaneous. The reaction takes several minutes on the average at ordinary temperatures.

Carbonic acid, because of its molecular structure, would be more informatively written as $(HO)_2CO$. It has a central carbon to which are attached two OH groups and one oxygen, looking like

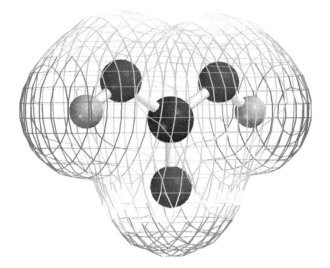

The shaded surface outlines the approximate size of the electron-containing space around the atoms' nuclei. The grid shades represent above average electron density (net negative charge: dark red) through below average (net positive charge: blue), respectively. The inner structure schematically indicates bonds among carbon (central black sphere), oxygen (dark red), and hydrogen (light grey).

The total solubility of carbon dioxide in *pure water* at 25°C (77°F) in contact with the *atmosphere*'s low 400 ppm pressure of carbon dioxide can be expressed as 0.0000136 moles per liter; ~0.6 mg per liter; one-tenth ounce per thousand gallons of water. Only about 0.1% of CO_2 ends up as H_2CO_3. Carbonic acid is present as 0.000000018 moles or about a million billion molecules per liter. A fraction of those molecules has ionized, releasing hydrogen ions, H^+, to the water. H^+ concentration is 0.0000025 moles per liter corresponding to a pH of 5.6. Rainwater typically has a pH of 5.2. Estimates are that rain removes 290 million tons of carbon dioxide from the atmosphere each year, a bit more than one-hundredth of 1% of the atmosphere's total carbon dioxide content.

Breakup, or dissociation, of carbonic acid in solution is written in the following chemists' short-hand notation:

$$H_2CO_3 \rightarrow H^+ \text{(aqueous)} + HCO_3^- \text{(aqueous)}$$

Besides H^+, the other product of this break-up is the negatively charged bicarbonate ion, HCO_3^-. This latter ion, as we have mentioned, can also behave as an acid (an even weaker acid than H_2CO_3), releasing the second H^+ according to

$$HCO_3^- \text{(aqueous)} \rightarrow H^+ \text{(aqueous)} + CO_3^{2-} \text{(aqueous)}$$

where CO_3^{2-} is the abbreviation for the *carbonate ion*. However, being a much, much weaker acid than the original carbonic acid, the bicarbonate ion contributes an insignificant amount of additional H^+. It does generate a very small quantity of carbonate amounting to about 5×10^{-11} (0.00000000005) moles per liter for water in contact with the sea level's atmosphere.

To summarize the behavior of carbon dioxide in otherwise pure water with an example: in a typical drop of water (about one-fortieth of a milliliter) there would be about 15 trillionths of a gram of carbon dioxide dissolved if the water were in equilibrium with ambient air. This corresponds to 200,000 billion molecules of carbon dioxide that have entered the drop, most becoming merely encapsulated within blankets of water molecules: CO_2(aqueous). Of the total, 260 billion nevertheless would be intact carbonic acid, H_2CO_3, as pictured previously; there would be 19 billion hydrogen ions and an equal number of bicarbonate ions ($\approx 0.01\%$ of the carbon dioxide total); there would be "only" 10,000 carbonate ions. Yet almost all of the drop of rainwater is water, H_2O: 800,000,000,000 billion molecules of water. These are still large numbers. If we considered a smaller amount of rainwater, a very tiny amount, such that the total *number* of carbon dioxide molecules was numerically equal to the population of Earth, 7 billion occupying a volume about that of a single plankton cell, then there would be 9 million carbonic acids, 660,000 hydrogen ions, another 660,000 bicarbonate ions, and one carbonate ion.

A schematic arrangement of 1,000 water molecules is shown on the left. On the right are represented more than three million molecules, the average number "surrounding" a single carbon dioxide molecule at equilibrium in the earlier description. Fifty times that count would show the amount of water relative to one carbonic acid molecule. These are dilute solutions indeed!

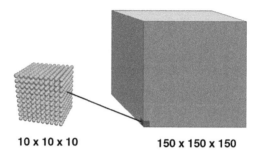

10 x 10 x 10   **150 x 150 x 150**

What makes this picture considerably more complicated is that the composition of the mixture we just described depends on temperature, on the amount of carbon dioxide in the air, and especially on the presence of other chemical species that might also be present in the water or in contact with the water.

TEMPERATURE EFFECTS

Generally, the solubility of a gas in water decreases as the water warms. A cold glass of water sitting out and warming up to room temperature can be seen to degas: to form bubbles. The bubbles contain air: nitrogen and oxygen. For carbon dioxide, if the temperature increases from 25°C to 30°C, the solubility of carbon dioxide *drops* about 13%. The graph indicates the decreasing solubility of carbon dioxide in water (on a weight-to-weight ratio basis: grams of CO_2 per gram of H_2O) as temperature increases, maintaining *one atmosphere* of pure carbon dioxide above the solution. (The same trend pertains to the lower pressure of carbon dioxide in the present day atmosphere, 400 ppm, reducing the vertical axis by a factor of 0.0004.) The point is, three times as much CO_2 can dissolve in frigid water (near the freezing point) as in tropical water in nature. Consequently, if you have information on the carbon dioxide content of seawater, you might have information on sea water temperature at the same atmospheric content of CO_2 without needing to resort to using a thermometer. A proxy.

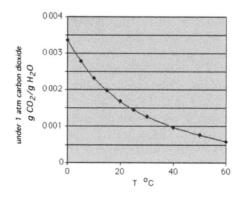

However, not only does the solubility of carbon dioxide in water vary with temperature, but so does the composition of the solution itself. We'll see examples of this incredibly important point shortly.

The solubility of carbon dioxide in ice is much smaller than in liquid water under ordinary circumstances. On the other hand, studies at high pressure, 10–20 atmospheres, show carbon dioxide can actually form a "stable hydrate." The structure of this complex arrangement, determined in 1999, was shown to correspond to an array of eight conjoined "cages" made from 46 linked water molecules. Each cage includes a trapped carbon dioxide so that the formula could be written* as $8CO_2 \cdot 46H_2O$. The hydrate retains all or most of its encapsulated carbon dioxide up to just about the normal melting point of ice. Often in the literature, the hydrate is symbolized more concisely but less precisely as $CO_2 \cdot 6H_2O$. Carbon dioxide hydrate has been hypothesized as a substance present in the polar regions of Mars.

pH EFFECTS

The relative amounts of the three dissolved carbon dioxide species ("carbonic acid," bicarbonate, and carbonate) in equilibrium with carbon dioxide in the atmosphere are shown here in the graph for various pH values of the solutions. With pure water, the equilibrium pH would be 5.6. Or, for one atmosphere carbon dioxide, it would be 4.0. At high pH values, that is basic (or *alkaline*) systems, all carbon dioxide varieties convert to the carbonate form, the doubly negatively charged ion CO_3^{2-}. At the opposite extreme, in strongly acid media (low pHs), everything is what we've labeled as "H_2CO_3" where the quotation marks are a reminder that only a tiny percentage of this is actually the H_2CO_3 molecule and the vast majority is carbon dioxide surrounded by, but not chemically bonded to, water molecules. At intermediate pH values, more or less near neutrality, the dominant form is the bicarbonate ion, HCO_3^-. In the figure, the thick curves track the relative composition as a function of pH of the aqueous medium when the temperature is 25°C (77°F). The thin lines do the same, but at 10°C (50°F), more representative of cool seawater, indicating that relative composition changes (slightly) with temperature as does solubility. These are *relative* changes. There is, of course, more carbon dioxide dissolved at 10°C than at 25°C as we noted in the preceding section.

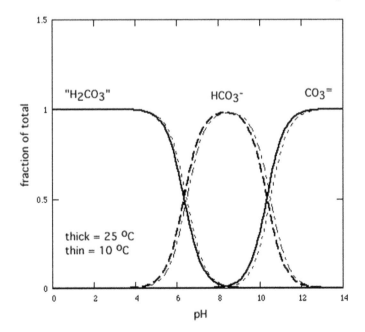

PRESSURE EFFECTS

The simplest way to start to look at the influence of gas pressure on solubility is to use an elegant law proposed in 1801 by William Henry (shown in the figure), a British physician and chemist. Henry's law says that the amount of gas dissolved in a non-reacting liquid is directly proportional to the pressure *of that gas* over the liquid.

Keeping the pressure of carbon dioxide constant but moderating the pressure of other gases does not alter the solubility of the CO_2 in the water. If we double the pressure of carbon dioxide, regardless of what we do to the other gases present in the air, we increase the amount dissolved. H_2O is a reacting liquid toward CO_2 since the mixture contains not only CO_2(aqueous) but also H_2CO_3, HCO_3^-, and CO_3^{2-}. Dissolving more carbon dioxide changes the pH, and keeping in mind carbon dioxide solubility is pH-dependent, we don't get a doubling of solubility. Conversely, if we release the pressure, as in opening a can of soda, or a bottle of champagne, or raising CO_2-saturated deeply-lying water to the surface as happened with tragic consequences at Lake Nyos, Africa (see Chapter 13), less gas is soluble and the excess bubbles out. A sealed can of carbonated soda is typically under a pressure of five atmospheres of carbon dioxide.

Another pressure effect should be mentioned briefly. That is, the effect on the composition of dissolved carbon dioxide species in water at various depths in water, that is, the consequences of changing water pressure. As it happens, there is only a very slight shift in the distribution among species with increasing pressure at kilometer depths, lessening the concentration of the CO_3^{2-} ion (if we ignore other features that vary with depth in the oceans such as bioactivity and also the complex chemical composition of ocean water).

CARBON DIOXIDE AND SEAWATER

Ocean water sits pretty much at a pH around 8.2. Following what we've seen, this indicates ocean waters did not simply evolve as pure water dissolving carbon dioxide and equilibrating, or the pH would currently be 5 to 6. From the graph below, you can see that the dominant form for dissolved carbon dioxide in the ocean at pH 8.2 must be the bicarbonate ion, HCO_3^-, with just a small contribution from carbonate and from undissociated carbonic acid, H_2CO_3.

The alkalinity of ocean water is mostly due to the bicarbonate. A rough measure of ocean alkalinity – its resistance to changing its pH – is the quantity of HCO_3^-, plus twice the amount of CO_3^{2-} present (as explained earlier), less any H^+ that is already present. This ignores the minor effects of other substances present in seawater (borates, hydroxide, phosphates, and silicates to mention a few) which also contribute to total alkalinity. Quantitatively and counter-intuitively, it has been shown that the total amount of carbon dioxide species dissolved, $CO_2(aqueous) + H_2CO_3 + HCO_3^- + CO_3^{2-}$, does not change much if the atmospheric level of carbon dioxide changes as long as the alkalinity of the ocean remains constant, which it essentially does. When the total amount of all dissolved carbon dioxide species remains unchanged though, this is not the same as saying the amounts of the different species are constant. Of special interest is the fact that, with constant alkalinity, increasing the amount of CO_2 in the air is accommodated by a reciprocal *decrease* of carbonate concentration in the ocean, shifting what was CO_3^{2-} into bicarbonate. An expanded discussion follows shortly. There are important real world consequences of this interdependence, among which is the necessity of carbonate for the growth of coral reefs.

Of considerable initial interest to us is whether or not there are any differences if – instead of pure water – we consider seawater or, for that matter, other mixtures of everyday practical composition. Is there any influence on carbon dioxide's behavior when other chemical species are present as well? The graph here compares the previous pH dependence of composition of carbon dioxide in *pure water* versus that in *seawater*, both at the same 10°C. The change is appreciable. Very briefly, we can discuss in what way and why this is so. When you look at that figure, note that at ocean pH, typically 8.1–8.3, there is a very significant difference for bicarbonate and also carbonate compared to pure water. These species are crucial to the understanding and behavior of incorporation of carbon dioxide into plankton, seashells, and minerals and their re-dissolution as well.

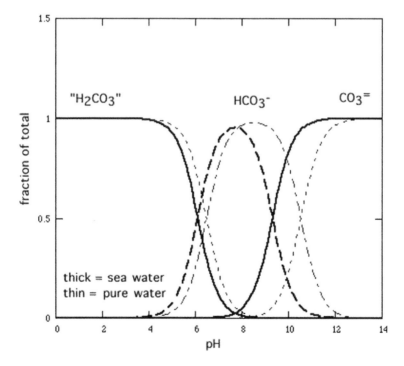

And even more complications occur in ocean water. For example, bicarbonate and carbonate exist not only as HCO_3^- and CO_3^{2-}, but also as the ions $CaHCO_3^+$ and $NaCO_3^-$ and as other species as well. These are "association ions" formed when Ca^{2+} and Na^+ combine (associate) with bicarbonate and carbonate, respectively. We have cited here just two of a large number of such "association complex" examples. A closer look at some of the influential carbonates will be the topic of Chapter 12.

RESTRAINT: THE REVELLE FACTOR

One hundred years ago, Svante Arrhenius argued that carbon dioxide levels were rising and that warming, due to the greenhouse effect, might be a concern (Chapter 6). Scientists at the time, however, were mostly dismissive of the idea for a very simple and logical reason. The ocean was known to have a capacity to absorb increasing quantities of carbon dioxide. Strictly speaking, this is correct…in the long run…at equilibrium. As far as carbon dioxide is concerned, however, equilibrium takes many centuries to achieve. On the shorter time scale of decades, the chemistry of atmospheric carbon dioxide exchange with ocean surface waters is surprisingly and counter-intuitively in the opposite direction of initial expectations. This was apparently first recognized in 1954 by American oceanographer Roger Revelle, pictured here on the left. The behavior is an acknowledgment of the surface ocean's carbon dioxide buffering effect, quantitatively embodied in what is now usually referred to as the Revelle (buffer) factor. Ordinarily, for gases such as nitrogen, oxygen, and argon, a 1% increase in their atmospheric pressure results in a 1% increase in the dissolved gas. (This is Henry's law, referred to earlier.) The *Revelle factor,* a measure of ocean buffering capacity, or its resistance to taking up more atmospheric carbon dioxide, is most simply put as

> the percent increase in atmospheric carbon dioxide pressure that brings about a 1% increase in the concentration total of all carbon dioxide species dissolving in ocean water.

The Revelle factor statement contains the key phrase "all carbon dioxide species" (also known as *dissolved inorganic carbon*) by which is meant "H_2CO_3" $+ HCO_3^- + CO_3^{2-}$. Henry's law involves only "H_2CO_3" or more accurately, CO_2(aqueous).

Not surprisingly, the value for this Revelle factor or ratio depends on many variables: temperature, alkalinity of the water, amount of dissolved carbonate species, and more. Low Revelle factors are associated with warm waters and high factors with cold waters. The figure on the right shows the global situation.* Values range between nine and 15, low values in tropical waters have a higher capacity for absorbing carbon dioxide from the atmosphere when the amount released into the atmosphere increases. At first glance, this seems backward for we have already noted that less carbon dioxide would dissolve at higher temperatures in pure water.

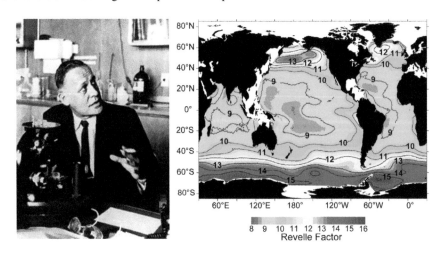

---

* C. L. Sabine et al., "The ocean's sink for anthropogenic CO_2," *Science* 305, 367 (2004).

When we considered the pressure effects of gases a few pages back, we introduced Henry's law that says, quite unequivocally, the amount of gas dissolved in a liquid is directly proportional to the pressure of the gas over the liquid. Doubling one doubles the other. If you have air above water in equilibrium with that water, some of the air will have dissolved in the water. Next double the air pressure, with a bicycle pump for example. After equilibrium is attained, you will have doubled the amount of oxygen and of nitrogen dissolved. And *vice versa* if you drop the pressure in half. Henry's law is the explanation of "the bends" that deep divers must avoid when rising rapidly from the high pressures associated with sea depths that dissolve nitrogen in the blood. Too sudden a decompression on ascending to the surface causes the nitrogen to bubble out in the blood and impede capillary circulation. For gases such as oxygen and nitrogen, the Revelle factor is 1: a 1% increase in atmospheric pressure of these gases brings about a 1% increase in the amount dissolved in water: simple and logical.

What to expect with increasing the carbon dioxide pressure in the atmosphere above seawater? Proportionately more CO_2 will dissolve. But unlike oxygen and nitrogen, which when dissolved are essentially the same chemical species but surrounded by water molecules, carbon dioxide can undergo chemical reactions in water. For pure water exposed to present day pressures of carbon dioxide in the atmosphere (400 ppm), the pH of the water at equilibrium would be about 5.6 (as we mentioned earlier). At that pH, essentially all the carbon dioxide is in the form of "H_2CO_3" as seen in an earlier graph. At ocean pH, though, because of the prevalent alkalinity, only a small fraction of the carbon dioxide species eludes reacting with water to form bicarbonate and carbonate. That small fraction remains in the form of solvated CO_2. As for oxygen and nitrogen, Henry's law refers specifically to equilibrium with this dissolved, but otherwise unchanged CO_2, not to the bicarbonate or carbonate species.

Increasing the amount of CO_2 – an anhydrous acid you may recall – in the atmosphere puts more acid into the ocean, dropping the pH. If you revisit the graph of how the various carbon dioxide species in water vary with pH, you'll notice that dropping the pH, moving to the left on the graph, results in lowering the relative CO_3^{2-} concentration. Expressing this in an alternative way, lowering the ocean pH causes the carbonate and bicarbonate amounts to shift and form "H_2CO_3," carbonic acid plus solvated, non-ionized CO_2. These shifts are very sensitive to small changes in H^+ concentration as in a buffered solution. In chemical symbolism, we are recognizing the importance of the reaction

$$CO_2 \left(\text{aqueous}\right) + H_2O + CO_3^{2-} \left(\text{aqueous}\right) \rightarrow 2HCO_3^- \left(\text{aqueous}\right)$$

Now we concede that two competing effects occur when increasing the carbon dioxide gas pressure in the atmosphere. At equilibrium, some of the change will be an increase in the amount dissolved due to higher gas pressure in the atmosphere. But some of the increase in CO_2(aqueous) will be due to internal rearrangement – resistance – within the water owing to the drop in pH and the accompanying shift in equilibrium away from bicarbonate and carbonate and back to forming CO_2 in the water. Carbonate and bicarbonate get recruited to supply CO_2(aqueous) necessary to comply with Henry's law that dictated how much is needed to equilibrate with the atmosphere.

The discussion staged here is a simplification of the work done by Revelle. Revelle used all of the chemical reactions occurring, generating an expression for the percent change in dissolved carbon dioxide species associated with a given percent change in pressure of carbon dioxide gas above the solution. Our version discussed earlier underplays the roles of various other characters in carbon dioxide buffering. These include borates, which contribute to pH sensitivity, and a variety of carbonate association complexes such as $CaHCO_3^+$ and several others which are intimately involved in carbon dioxide's many hiding places. The Revelle factor, when fully considered, tells us that a 50% increase in carbon dioxide in the atmosphere typically results in only a several percent increase of carbon dioxide in all its forms in the water phase, that is, *dissolved inorganic carbon* (DIC).

If we increase the amount of carbon dioxide in the atmosphere from recent levels to nearly 50% higher, the composition of carbon dioxide species in seawater undergoes the changes laid out in the table. The values are in micromoles* per liter.

Species	Ocean: 380 ppm CO_2 in Atmosphere	Ocean: 560 ppm CO_2 in Atmosphere	Percent Change
"H_2CO_3,"	13	18	+38%
HCO_3^-	1827	1925	+ 5%
CO_3^{2-}	186	146	−22%
Total dissolved	2026	2090	+3.2

From Kleypas, J. A. et al. "Impacts of Ocean Acidification on Coral Reefs and Other Marine Calcifers," NSF, NOAA, USGS (2006).

Note that "H_2CO_3," which is almost entirely CO_2(aqueous), increases by about 38% as is close to what is expected from Henry's law. But total dissolved inorganic carbon ("H_2CO_3" + HCO_3^- + CO_3^{2-}) increases from 0.002026 to 0.002090 moles per liter, changing only 3.2% upon a 47% increase in the atmospheric carbon dioxide total. Most of the dissolved carbon dioxide is in the form of the bicarbonate ion, about 90% of the totals. An increase in hydrogen ion concentration also accompanies the addition of acidic carbon dioxide, dropping the pH from 8.05 to 7.91. Also occurring is a decrease in carbonate by 22%. Putting more carbon dioxide into the ocean decreases carbonate, an ion required, as has been noted, for the growth of coral. A level of 280 ppm (0.028%) CO_2 corresponds to what several studies reasonably imply was the atmospheric situation for more than 1,000 years prior to the Industrial Age that commenced around 1850–1860 and also seems to have been the level prior to 30 My ago. (See Chapter 8.) At the very least, you can see from this discussion that the addition of carbon dioxide drops the pH: the solution becomes more acidic. You would be correct in inferring that the reverse is true, a lower carbon dioxide level would be associated with a higher ocean pH. (Any pH-dependent process is thus potentially a proxy for CO_2 atmospheric levels.) Specific to the numbers in the table, a 3.5% change in the total of all carbon dioxide species would be brought about by a 50% increase in the carbon dioxide atmospheric pressure. The results imply that a 1% change in all dissolved carbon dioxide species would be associated with a 14.3% increase in atmospheric carbon dioxide or a Revelle factor of 14.3.

Ultimately, at equilibrium, most of any additional atmospheric carbon dioxide could dissolve in the oceans, mixing with the vast quantities of colder water under enormous pressure at greater depths than near the surface. Moreover, the acid would react with sedimented solid calcium carbonate to produce the bicarbonate ion (and the calcium ion), effectuating a buffering of the drop in pH. However, we are talking about thousands of years needed before a new equilibrium is achieved.

FRESH WATER CONSIDERATIONS

Significant variations from oceanic behavior emerge when considering instead freshwater systems. Less than 3% of Earth's crustal waters are not in the oceans. Underground springs or wells offer some interesting situations. Carbon dioxide concentrations in such natural freshwaters are often 10-100 times greater than in pure water equilibrated with atmospheric CO_2. In one particular spring water supply (at 10°C) there was 0.45% carbon dioxide gas *versus* 0.035% carbon dioxide in the atmosphere. For some representative well waters that were studied, values like 1.74% (well

---

* Micromole = millionths of a mole.

temperature 14°C) and another at 2.00% (11.8°C) were observed. These waters are usually found to be in contact with chemical species such as calcium carbonate as opposed to contact with the atmosphere.

Air entrapped in soil is also significantly enriched in carbon dioxide, a fact that can contribute to the carbon dioxide enrichments of freshwaters noted earlier. Representative values from a study on the island of Trinidad are shown in the table here from H. D. Holland's *The Chemistry of the Atmosphere and Oceans*. Keep in mind that 1% corresponds to 10,000 ppm.

Depth	% CO_2 Trapped in Air during Dry Months	% CO_2 Trapped in Air during Wet Months
10 cm	1.5	6.5
25 cm	1.2	8.5
120 cm	5.1	9.6

Carbon dioxide trapped in underground soil most likely comes from root respiration and from the decay of plant matter brought about by microbial action. The global amount of organic carbon in soils is equivalent to about 11,000 Gt of carbon dioxide, nearly three times the atmospheric total.* Respiration by microbes, fungi, roots, and invertebrates is estimated to release less than 300 Gt per year. Of course, oxygen is required. A major portion of soil carbon dioxide slowly diffuses back into the atmosphere. Estimates of the amount that works downward into groundwater where it might remain for hundreds or even thousands of years is about a gigaton per year. Local values are sensitive to soil pH which, in turn, depends on bedrock type and climate (temperature and rainfall).

All in all, this chapter demonstrates that both macro and micro perspectives are needed in order to come to terms with the complexity of carbon dioxide's behavior.

---

* C. E. Hicks Pries, C. Castanha, R. C. Porras, and M. S. Torn, "The whole-soil carbon flux in response to warming," *Science* 355, 1420–1423 (2017).

11 Going with the Flow

Time is a sort of river of passing events, and strong is its current.

MARCUS AURELIUS ANTONINUS

Carbon dioxide, being soluble in water, appears in lakes, streams, rain, and so forth, sometimes in surprisingly large quantities. Not unexpectedly, a close look at some of the interplay between CO_2 and various natural bodies of water is central to our explorations.

RIVERS AND OCEANS

Most of the CO_2 in rivers and oceans is in the form of bicarbonate. The amount of bicarbonate (HCO_3^-) in river waters averages around 40–60 ppm (mg/kg). Compare this with the ocean where the average is 1,800 mg/kg. From Chapter 10, 0.6 ppm is carbon dioxide's solubility in *pure* water at 25°C at equilibrium with the atmosphere mostly as CO_2(aqueous). Compare those values with that for the sodium ion, Na^+. Its average level is 7 mg/kg in typical river water yet 10,800 mg/kg in oceans and nearly nine times higher yet in the Dead Sea. In contrast to runoff waters trapped in the Dead Sea – a cul-de-sac for the Jordan River – salty oceans are not the result of river water simply becoming more concentrated due to the loss of water by evaporation.

The total amount of sodium in rivers and oceans can be obtained from the total mass of water in these bodies. A comparison among these is telling.

Oceans contain 1.4×10^{21} kg (1.4 billion billion tons or 1.4 billion km^3) seawater and their average depth is 4 km. That seems like quite a load, but it is only 0.025% of Earth's mass. Rivers flow with a total rate of 4.6×10^{16} kg water per year: a volume of 50,000 km^3 per year. The refill time of the ocean, hypothetically, can then be gotten just from the ratio of these two values which is 1.4×10^{21} kg $\div 4.6 \times 10^{16}$ kg per year or 30,000 years. This can be viewed as an average residence time for water in the ocean. Cycling time of ocean water currents that are part of the intricate circulation pattern is less than 1,000 years. The remainder, by comparison, could be considered almost stagnant. Similarly, using the previous values for sodium, the oceans contain 10,800 mg/kg·(1.4×10^{21} kg) = 1.5×10^{19} kg sodium. The rivers contribute 7 mg/kg·(4.6×10^{16} kg per year) = 3.2×10^{11} kg per year; the ratio of these two gives an accumulation time for Na^+ as 47 million years. For bicarbonate the residence time works out to be about 80,000 years. These time intervals are not contradictory but merely conceal the underlying complexity of how various chemical species are transported.

WATER CARRIER

Carbon dioxide transport is among the most complicated behaviors to monitor. Sodium is easier. Na^+ doesn't readily return to resupply the rivers (47 million years). Water itself is recycling (every 30,000 years). Bicarbonate (HCO_3^-) recycles more slowly (80,000 years). In effect, sodium does not recycle. The significance of these residence times is not simple though. Their interpretation is fraught with complications. Despite the previous statement about the permanence of marine Na^+, even sodium is slightly cycled back to the river supplies through sea-borne aerosols and brackish water for example. And, of course, there are previously deposited evaporated brines – ancient salt deposits – that re-dissolve in water and are carried by the river supply. If we knew the history of all those salt forms, the numbers should reconcile perfectly. However, we're in the opposite situation and need to synthesize the complete history from the current evidence.

Water circulation is a key contributor to the fate of carbon dioxide on Earth's surface. The rates at which rivers deposit substances, including carbon dioxide and its derivatives, into the oceans varies, undoubtedly, with the flow rate of individual rivers. More erosion takes place at swifter flow and with higher volumes of water. The world map here shows the geographical variation of annual erosion rates in metric tons (1,000 kilograms) per square kilometer. These are averages and vary seasonally and from year to year.

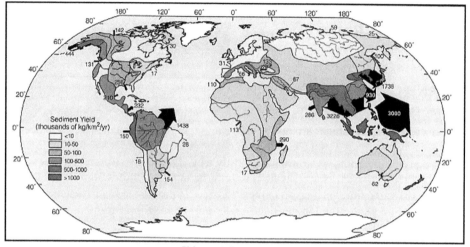

Skinner & Porter (1994)

The erosion is caused by surface runoff of water. Units of runoff (not erosion) are the volume of water per year that drains a chosen area of surface. For discussion of runoff, a convenient volume is cm^3 and the unit area is cm^2 so that runoff is expressed as cm^3/yr per cm^2 or $cm^3/yr \cdot cm^2$ corresponding to cm/yr. This could be viewed as the depth of coverage by flowing water averaged over a year if everything were uniform. These are strange units to picture. So in the interest of clarity, since a cm^3 is the same as a mL, runoff will be expressed here as mL/yr per cm^2. The world average for runoff is about 30 mL/yr per cm^2. This is identical to 12 inches of annual rainfall or 200 million gallons/year per square mile. That is an impressive amount of transport facilitating carbon dioxide's journeys.

There are significant departures from average runoff. The Amazon River has well over 1,000 tributaries. It drains half the continent of South America and at its mouth is 150 miles wide and 300 feet deep. The Amazon carries about one-sixth of the total water in rivers and, not surprisingly, has a very high runoff: ~85 mL/yr per cm^2, about three times the world average. Despite this, its dissolved bicarbonate content is only ~15 ppm, well below the world average of 40–60 ppm. The explanation is that much of the rainfall runoff region is underlain with highly leached sediments that do not contribute much bicarbonate to the draining water. In contrast, water saturated with calcite ($CaCO_3$) should have a bicarbonate concentration of ~70 ppm in otherwise clean situations.

Streams and rivers frequently have an excess of carbon dioxide compared to the amount equilibrated with the atmosphere. In such circumstances they can be considered sources of CO_2, returning in excess of 2 Gt annually to the atmosphere.*

A study of the St. Lawrence River and its tributaries in eastern Canada supports this explanation. Tributaries north of the main trunk of the St. Lawrence contain about 12 ppm bicarbonate and those from southern tributaries transport a robust 114 ppm. (Calcium ion concentrations in the waters follow a similar partition.) Both values are consistent with known deposits of calcite and dolomite carbonates underlying the surface south of the river.

An even more dramatic example is on the Pacific coast of Mexico where the Rio Ameca drains over igneous and metamorphic rocks starting near Guadalajara and draining into the ocean near Puerto Vallarta. The river water has as much as 375 ppm bicarbonate. The CO_2 pressure in the river water actually exceeds that of CO_2 in the atmosphere. The lack of equilibrium persists because the loss of the gas to the atmosphere cannot keep up with the feed from ground waters.

Minerals are not the only source of carbon dioxide and bicarbonate in river waters. If there is rich growth of vegetation, plant respiration – consumption of oxygen – can contribute carbon dioxide into runoff waters. So can oxidation of elemental carbon or organic material in rocks. It goes without saying that the atmosphere contributes to CO_2 concentration in dilute waters through seeking equilibrium levels of dissolved gas.

---

* P. A. Raymond et al., "Global carbon dioxide emissions from inland waters," *Nature* 503, 355 (2013).

THE OCEANS

Oceans occupy just over 70% of Earth's surface. Oceans contain all water save the 3% of "fresh" water (including snow and ice). Although the average ocean depth is 4 km, this isn't much compared to Earth's overall dimensions. Picture a stack of three reams of paper (500 sheets per ream). The ocean's average depth is one sheet thick on this scale.

Ocean bulk waters are layered according to their density. The figure here shows how seawater density would vary with depth if the water were stagnant. In the figure, "Pycnocline" refers to the approximate depth at which the change in density* is sharpest. It acts as a barrier to vertical water exchange. Temperature differences may be as great as 20°C. But much seawater is not stationary. Water flows away from polar regions, heading north or south toward the equator, along constant density streams. These flows don't mix too much with surrounding waters.

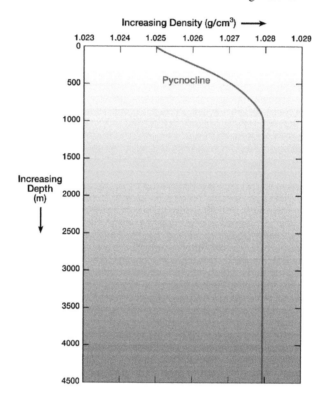

Polar waters are cold. Water is most dense near its freezing point. Stating that polar waters are dense means a given volume is heavier compared to the same volume of warm water. Almost all deep waters are cold because, being denser, they sink, like a metal coin compared to a wooden coin. If initially near the surface, cold water picks up carbon dioxide from the atmosphere. The colder the water, the greater is the solubility of carbon dioxide at equilibrium. The amount exchanged between the atmosphere and the surface water depends on how long the water is exposed and on the buffering capacity of that surface water (see Chapter 10), both of which depend on a number of other factors. A map of the CO_2 exchange between air and sea surface is shown here. Coastal results are in dark colors and open seas are shown in light colors. Both domains range from blue, for CO_2 sinks, to red, for sources vis-à-vis the atmosphere.[†]

---

* The Greek word for dense or thick is pyknos (πυκνοσ).
[†] N. Gruber, "Ocean biogeochemistry: Carbon at the coastal interface," *Nature* 517, 148 (2015).

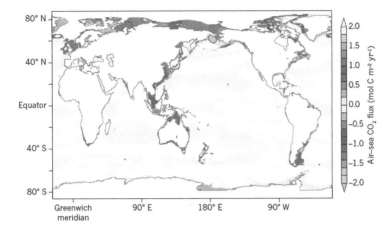

As cold, dense surface waters move away, they sink to greater depths, as they must. During the subduction to greater depth, more carbon dioxide is added to the flowing volume, not from the atmosphere (which is out of contact), but from oxidation of sunken biological debris and from other local sources. For example, bicarbonate extracted from seawater and converted into insoluble carbonates for shells and *tests* (from Latin *testa*=shell) of dinoflagellates and sea urchins (counter-intuitively) *release* carbon dioxide as part of the conversion process.

$$Ca^{2+} + 2\,HCO_3^- \rightarrow CaCO_3 + H_2O + CO_2$$

Such mineralization is the reverse of carbonate weathering, a topic embraced in Chapter 16.

Ocean acidity is buffered at approximately pH 8.2, mostly through the role played by bicarbonate. Putting carbon dioxide into the ocean drops the pH. If the pH drops, making the ocean waters more acidic, calcium carbonate deposits in contact with the more acidic waters begin to dissolve reversing the previous chemical reaction, consuming CO_2. This, in turn, explains the greater capacity of the ocean for carbon dioxide. Why? We dwelled on this in Chapter 10, "Carbon Dioxide and Water." Dissolution of sedimentary carbonates will ameliorate recent increases of carbon dioxide in the atmosphere. But that's a long-term perspective. One estimate is that only 7 Gt of carbon dioxide per year is being sequestered by this route. Ocean storage takes time to brandish its full effectiveness. The atmosphere contains 3,200 Gt of carbon dioxide. Annual exchange with the ocean is several percent of this.

As has been emphasized, most of the carbon dioxide in the ocean is in the form of bicarbonate, HCO_3^-. Aside from the polar regions, warm surface waters (relatively warm, that is) down to depths of perhaps 70 m contain 2,200 Gt of carbon dioxide in its alternative form, bicarbonate. (Observe how close this is to atmospheric content noted in the previous paragraph.) Dropping to greater depths, the water temperature drops as well, sometimes quite sharply. The boundary layer itself between "warm" and "cold" is termed the "thermocline." The thermocline is the layer where the temperature change is most rapid just as the pycnocline addressed density changes. Down to the thermocline at 1,000 m, seawater contains 35,000 Gt of carbon dioxide equivalent. The deep ocean below the thermocline has 110,000 Gt, nearly three-fourths of the ocean total.

Cold water, originating in the typically frigid North Atlantic, is dense and sinks. In the Denmark Strait, between Iceland and Greenland, ocean water plunges 3.5 km delivering a flow moving beneath the ocean surface at 1.4 m per second (~3 mph), a flow that is a current 25 times greater than that of the Amazon River: 5 million cubic meters (more than 1 billion gallons) *versus* 0.2 million cubic meters for the Amazon. That's each second! And the Amazon carries one-sixth of the world's river total.

Of course, the Denmark Strait subsurface water must go someplace, pushing existing water out of its way. So we have a flow at great depths that heads almost due south toward the Antarctic, eventually rounding South Africa and swinging north again as it warms up through the Indian and Pacific Oceans toward the North Pacific. Some of the warmer, less dense flow rises to shallow depths and loops back across the tropical Pacific past Australia, then south of Africa and back along the western Atlantic to the North Atlantic once more. The three-dimensional figure shown here does an excellent, if concise, job of picturing this infraoceanic river.

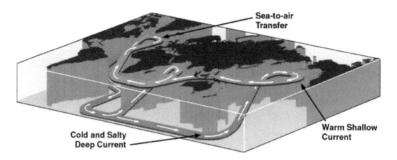

If all this water flow does not mix with the surrounding ocean waters, its ratio of radiocarbon (^{14}C) to the common, natural ^{12}C can determine how long the water has been isolated from the atmosphere where the radiocarbon is produced. The flow in the Atlantic has a ratio corresponding to 90% of the atmospheric ratio. The 10% loss is by radioactive decay. Since the carbon isotopes behave essentially the same in their chemistry and physics, but the ^{14}C is fading from existence with time elapsing, this ratio signifies an average age of the water carrying the radiocarbon of 800 years (!). (See the second figure in Chapter 5.) This assumes, fairly reasonably, that there is no exchange of the water flow with older, nearly stagnant, surrounding volumes. In the Pacific, the deep water ratio is lower still – 77% – from which an "age" of 2,100 years can be inferred. From these two "ages," we can get something like an average residence time of carbon dioxide in ocean water streams as 1,500 years. The total carbon dioxide content, if we pretend this is its residence time, ignoring the large relatively stationary volume, then turns over at a rate of 1.5×10^{17} kg per 1,500 years or just under 100 Gt carbon dioxide per year, a bit more than 3% of the atmosphere's total. A 2012 mapping* of radiocarbon shows the age distribution in detail.

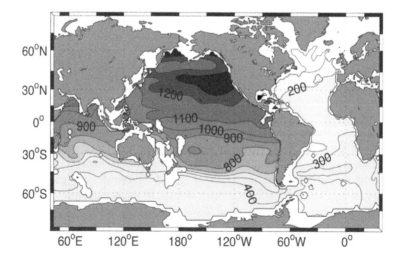

---

* G. Gebbia and P. Huybers, "The mean age of ocean waters inferred from radiocarbon observations: Sensitivity to surface sources and accounting for mixing histories," *J. Phys. Oceanogr.* 42, 291 (2012).

FEEDING THE OCEANS

The Niagara River, between Canada and the United States, supplying the spectacular Niagara Falls, carries a mere 200 km^3 per year (about 1% of the Great Lakes total volume), much of it utilized to generate in excess of 4 million kilowatts of electric power at the base of the 70 m drop. The river is a central component of the waterway that connects the Great Lakes to the Saint Lawrence Seaway that empties into the North Atlantic by the Canadian Maritime Provinces.

Elsewhere, annual discharge of the six largest rivers in the Eurasian continental landmass emptying into the Arctic Ocean has been monitored over six decades. These rivers (Kolyma, Lena, Ob, Pechora, Severnaya Drina, and Yenisey) drain about 2,000 km^3 of river water each year from most of the landmass in the Eurasian Arctic. Despite difficulties in determining long-term trends in individual rivers, a cumulative 7% increase in runoff over the last 60 years is apparent. The significance is that a large influx of fresh water into the surfaces of the North Atlantic, with which the Arctic connects, decreases seawater salinity and density, thereby potentially disrupting the ocean's "thermohaline" circulation pattern from its current blueprint, known appropriately as its "conveyer belt."

The ocean convections, at least now, carry high pH (8.2) and high carbonate concentrations in the North Atlantic down to great depths. In contrast to this behavior, the Pacific is "unventilated," not exchanging contents with strata upward or downward, leaving its carbonate concentration and pH lower despite the higher concentration at great depths, for example at 3,000 m shown here in the figure. At these depths, the increase of about 10% traveling from the North Atlantic to the North Pacific is understood to be due to oxidation of organic matter in the deep ocean.

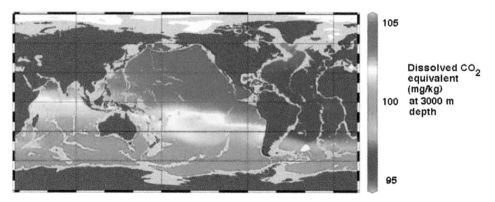

Dissolved CO_2 equivalent (mg/kg) at 3000 m depth

There is much evidence that the interplay between pH and carbon dioxide cycling involves the transitions between glacial and interglacial periods including, most interestingly, the ice ages. Not surprisingly, there is considerable interest in the past distribution of carbon dioxide throughout the oceans.

Most of the sea bottom terrain is plain-like: flat with occasional rolling hills. The terrain is covered by mud consisting of carbonate and siliceous (containing *silica*, silicon dioxide) oozes kilometers deep made from microscopic remains of *foraminiferans*, *diatoms*, and *radiolarians* that died and settled over the ages. The archives of some of what we know of the oceans' past can be read from microscopic shells of *foraminifera*, that is, plankton. Their shells are calcium carbonate. Some species thrive in warm waters, some in cold waters. Both the composition of the shells and the distribution of the plankton skeletons reveal much about temperature, chemistry, and circulation in the ocean (as will be discussed in Chapter 12, "Carbonates: The Enduring Legacy"). However, high pressure at ocean depths increases the solubility of calcium carbonate. Skeletons are then understandably degraded at depths and not profusely found. Shells are mostly unaffected if they settle on high undersea ridges which is where useful data are retrieved.

DEEP SEA VENTS

Subduction of Earth's tectonic plates has been mentioned previously, and especially with regard to volcanoes (Chapter 13). In subduction, crustal plates are colliding and one of the plates is forced downward, plunging under the other. The reverse of this can occur as well, rending the crust apart with plates separating. Such action is known to take place under the oceans in several locations. When submarine ocean waters touch the exposed hot magma, hydrothermal events transpire, much like the geysers so well-known on the terrestrial surface as at Yellowstone, for example. The submarine geysers are also known as volcanoes of the deep, or deep water seeps, but most commonly, as deep sea vents.

Deep sea vents were discovered first near the Galápagos off the coast of Ecuador in 1977. There are now 25 known locations with more than 500 active vents.

Greater than 75% of Earth's volcanic activity takes place below the sea surface, a fact that has been known for less than 50 years. Hydrothermal forces are so great at mid-ocean ridges that 1.5×10^{14} kg of ocean water is circulated through hot volcanic crust each year. This corresponds to the entire volume of global marine waters recycling through the ridges in ~10 million years (not that they actually do). The cycle is an important source of calcium for ocean waters and calcium, the fifth most abundant element after O, Si, Al, and Fe, is a pivotal ion in the chemistry of carbon dioxide. Carbon dioxide itself is also emitted at these sub-oceanic volcanoes, the production estimated to be 37 million tons per year. CO_2 from sea vents does not end up in the atmosphere for ages and marine waters around deep sea vents can be highly and atypically enriched in carbon dioxide.

In analogy to *photosynthesis*, that is, like the reaction in plants producing carbohydrates (abbreviated as $[CH_2O]$)

$$CO_2 + H_2O \rightarrow [CH_2O] + O_2$$

there is a similar process, sometimes referred to as *chemosynthesis*, which generates organic material using hydrosulfide, HS^-, as the energy source rather than light (hence *chemo* rather than *photo*). At hydrothermal vents, sulfide, an ionic form of sulfur, is produced by the reaction of seawater with hot rock deep within the ocean crust. Hydrogen sulfide, H_2S, the same gas that gives rotten eggs its characteristic stench, is mildly acidic. In sea water, it becomes hydrosulfide and performs chemosynthesis according to the (over-simplified) expression

$$HS^- + CO_2 + O_2 + H_2O \rightarrow [CH_2O] + HSO_4^-$$

HSO_4^- is the hydrogen sulfate ion (also known as bisulfate). As in the case of photosynthesis (Chapter 14), the reaction symbolized here is actually a much more intricate sequence of many separate biochemical steps. The enzyme-enabled carbon dioxide fixation cycle used by many, but not all, chemosynthetic bacteria is nearly identical to the so-called Calvin cycle used by photosynthetic plants (Chapter 14).

The visible spectacle of microbial activity at deep sea hydrothermal vents is the transient "blooms" of bacteria that accompany subsurface volcanic eruptions. Soon after the 1991 eruption on the East Pacific Rise, west of the Galápagos and south of Acapulco, submersible divers observed a newly deposited 5 cm blanket of white flocculent material over an area several kilometers in length and 50–100 m wide. The volume was sufficient to generate "white-out" conditions. The bloom was allegedly the sulfide-oxidizing primitive microbe *Archaea*. Such blooms are supposedly fed by enriched volatile nutrients such as carbon dioxide and hydrogen sulfide from the undersea eruptions. Carbon dioxide enrichment is comparable to that in carbonated beverages. That East Pacific Rise is site of one of the fastest spreading sections of the mid-ocean ridge, with crust spreading velocities found to be as high as 120 mm (~5 inches) per year, or 75 miles per million years.

In some cases, sea vent living species may date their evolution back several hundred million years, to the Paleozoic Era. The giant tubeworm, for instance, first discovered in the late 1970s, has some extremely unusual features, especially for an invertebrate that can be up to 2 m in length. It has no mouth. And it has no digestive system. Instead, it has a specialized organ, the *trophosome*, ("trophe" is Greek for nourishment) deep inside the animal. The trophosome is 15% of the worm's body weight. Each gram of the trophosome tissue contains several trillion symbiotic bacteria.

The usual flow of carbon dioxide, an end product of metabolism (respiration), is out of an animal. But symbionts, the bacteria mutually beneficial to the tubeworm, require a *net uptake* of carbon dioxide *into* the microbe for chemosynthesis just as plants require CO_2 for photosynthesis. The legacy of carbon dioxide has its freakish aspects.

Seawater contains plenty of bicarbonate ions but is not a directly useful source of CO_2 to the symbionts. The electric charge on the bicarbonate ions hinders them from diffusing into the tissue. CO_2 is present in small amounts, freely diffusible, and taken up by the symbionts for carbon fixation. As noted in Chapter 10, as pH decreases, bicarbonate concentrations decrease and carbon dioxide increases (and *vice versa*). At deep sea vents, the locally lower pH (~6) and enrichment of carbon dioxide from the eruptions elevates the concentration of diffusible carbon dioxide by a factor of about a thousand compared with ordinary deep seawater. This is beneficial to the symbiont bacteria's uptake of carbon dioxide. Uptake is further aided by the higher pH (7.3–7.4) of the tubeworm blood which facilitates the conversion of CO_2 back to bicarbonate, maintaining a diffusion-aiding CO_2 deficit on the "inside" relative to the "outside."

We'll finish this deep sea journey by noting that there are other fascinating aspects to the tubeworms' adaptation to their peculiar environment. Warm (15°C) hydrothermal vent water contains high sulfide concentrations but barely any oxygen. Oxygen is needed for the demands of the metabolism of the host and symbiont including oxidation of sulfide. In contrast, cold (2°C) seawater contains oxygen, but little sulfide. The worms live at the interface between the vent water and nearly frigid seawater which mix turbulently and bathe the worm in fluctuating temperature environments and alternating (sulfide *versus* oxygen) feed stocks on time scales of several seconds. They seem to do okay.

12 Carbonates
The Enduring Legacy

A great chapter of the history of the world is written in chalk.

THOMAS HENRY HUXLEY

SETTING THE SCENE

Starting at a depth of 5 km and for the next 85 km deeper into Earth's skin we have what is called the crust, structured as a jigsaw of huge plates. The crust is lean below the oceans, a mere 6 km thick. At the other extreme, it is especially robust at the Himalayas underneath which are 70 km of crust. But the crust contains just 0.7% of Earth's mass. So even though it is important and influential to our carbon dioxide legacy, the crust is a small piece of the pie. Earth's crust, the armor of plates, is very dynamic, being altered by earthquakes, erosion, and subduction, and is regenerated by volcanoes and at spreading sub-ocean ridges.

The next zone down from the crust is called the mantle. It extends halfway to the center of Earth where pressure exceeds a million atmospheres. The mantle contains two-thirds of Earth's mass with temperatures reaching thousands of degrees. Subduction is the sinking of crustal plates into the mantle and generates magma by melting rock.

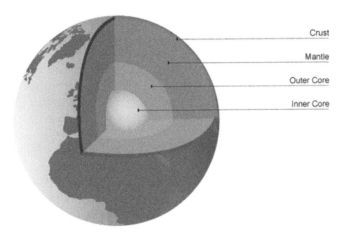

Crust

Mantle

Outer Core

Inner Core

WHITHER THE OCEAN'S CARBON DIOXIDE?

The ~3,200 billion tons of carbon dioxide in the atmosphere represent a tiny fraction of the carbon that is found in Earth's crust: ~20,000 *trillion* tons in all. Much of this carbon was carbon dioxide at some point in time and a good portion of it is still tied up as carbonate, CO_3^{2-}, in chalk, dolomite, limestone, and many other deposited materials, much of it sediment. All natural bodies of water sit on sediment. Additionally, about 75% of all land is underlain with sedimentary rock. Sediment is essentially recycled detritus and most often stratified. Strata – the layers – have thicknesses that can be as large as 8 to 10 miles. Most carbonates produced in the last half-billion years are skeletons of animals whose structure and composition took advantage of the chemistry of the ocean. Some examples of these skeletons are shown here. (Eggshells, for instance, are also calcium carbonate.)

Despite all this locked-up carbon dioxide, it is estimated that nearly half of the mass of the mantle consists of compounds of carbon dioxide's relative from the Periodic Table (see Chapter 1), silicon dioxide (SiO_2), through "silicates," an analog of carbonates. As we will see in Chapter 16, "Weathering," there is interplay between carbonates and silicates that profoundly affects the travels and travails of carbon dioxide.

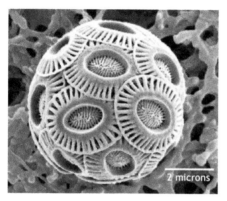 

Evidence implies that it is only during the most recent half-billion years – the Phanerozoic Era – that life forms developed skeletons.

Globally, the total mass of limestone, a prime example of sedimentary material, is ~350,000 billion tons. Living or formerly living organisms are responsible for nearly all calcium carbonate precipitation in the ocean today! The skeletal structures are abundant. Nevertheless, this issue is not totally one-sided because there is plenty of evidence of inorganic processes, not involving living organisms, leading to aragonite precipitation. But one-quarter of this is arguably Precambrian arising from simple chemical precipitation of carbonate with calcium ions. Calcium, element number 20, is the fifth most abundant element in Earth's crust. (The Cambrian was the most ancient period in the Phanerozoic Era.) The emergence of biomineralizers, life forms that generate solid material from dissolved substances, like those whose skeletons are shown here, made sedimentation significant.

Solid $CaCO_3$ exists predominantly in two forms, *calcite* and *aragonite*. Both these minerals are almost three times as dense as water, the aragonite being about 8% denser still than the calcite. A *mineral* is a naturally occurring, inorganic, homogenous solid having a definite chemical composition and characteristic crystalline properties. *Rock* is a naturally formed mixture of minerals. Other carbonate minerals of interest include dolomite (calcium magnesium carbonate), siderite (iron carbonate), ankerite (calcium magnesium iron carbonate), rhodochrosite, grapestone, and oolite. Nacre, pearls, and abalone shells are up to 95% calcium carbonate. There are hundreds of carbonates occurring in nature. Interestingly, both calcite and dolomite have recently been detected in the dust around distant stars.*

CHALK

Seventy million years ago, the east coast of England was below sea level. Algae skeletons, *coccoliths*, rich in calcium carbonate settled as a mud, slowly accumulating at perhaps a fraction of a millimeter every year. The previous picture on the left is one of the coccolith-bearing algae, a species of *phytoplankton*. The circular doily-like armor patches are the coccoliths formed from calcium carbonate secretions by the living algae. Phytoplankton are microscopic organisms that

---

* F. Kemper et al., "Detection of carbonates in dust shells around evolved stars," *Nature*, 415, 295–7 (2002).

live and drift in the sunlit upper layer of the oceans. There are about 5,000 different species of phytoplankton known. Astonishingly, their total live mass exceeds that of all other marine animals combined, including fish and whales. Over time, the settled accumulation of skeletons of these one-celled microscopic creatures built up as thick as 500 m in places including what is now the English Channel (La Manche in French and Het Kanaal in Dutch). Fossils of other sea creatures are occasionally deposited in with the layers. At some point, arguably 25–30 million years ago, the sea level changed as a result of crustal motions, and we have the spectacular 100-meter-high white cliffs of Dover to admire.

The white (chalk) cliffs of Dover, England

Thomas Henry Huxley, the well-known English philosopher and scientist of the nineteenth century, was one of the founders of the scientific method. His 1868 essay on "a piece of chalk" is masterful in its clarity, organization, and critical logic. It also bears on our subject. A transcription follows in the Appendix.

An important aspect of phytoplankton and their calcium carbonate armor structures is their role played in the carbon dioxide legacy. These algae are found in all marine environments except polar. Under conditions favorable to the plankton, extensive blooms can occur, sometimes covering an ocean area greater than that of Iceland. Moreover, the algae in the blooms seem to produce an excess capacity of armor pieces, the coccoliths made of nearly pure calcium carbonate. All of these float near the surface reflecting a large portion of sunlight that would otherwise warm the water, turning the ocean color into a milky turquoise. Between their influence on albedo (reflecting power) and their eventual sinking to the ocean floor, the phytoplankton blooms can cause unusual disruptions to the ocean's chemistry and in the chemical forms carbon dioxide will assume. An understanding of carbonate solubility is critical to understanding the many processes.

SOLUBILITY, SATURATION, AND SUPERSATURATION

The solubility of $CaCO_3$ as calcite in otherwise pure water at 25°C is 0.00575 g per liter. (A gram per liter can be viewed as an ounce in a gallon.) Aragonite is 23% more soluble. Vaterite is twice as soluble and hydrocalcite, a hydrated form of calcite, $CaCO_3 \cdot H_2O$, is almost three times as soluble. Vaterite is a form of $CaCO_3$ that tends to generate micron-sized spherical deposits. It is most familiar as one form of gallstones.

The solubility of calcium carbonate varies with temperature as shown here. We point out the somewhat unusual fact that $CaCO_3$ becomes less soluble in warmer water. That is a characteristic we will see plays a part in revealing CO_2 interaction with the ocean, a topic we introduced in the last chapter.

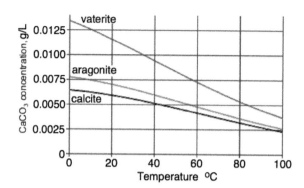

Most surface waters of the ocean are "supersaturated" with respect to calcium carbonate. That is, they actually contain more dissolved calcium carbonate – as calcium ions and carbonate ions – than they would at equilibrium. In contrast, for reasons that are well understood, deep ocean waters have the capability to dissolve more $CaCO_3$ than they presently contain: they are unsaturated. The reasons for the latter situation include decreasing temperature and increasing pressure with depth both of which foster dissolution. Additionally, CO_2 from microbe decomposition of falling organic debris increases acidity, dissolving $CaCO_3$. A result of unsaturation of calcium carbonate in deep water is the dissolution of calcite and aragonite deposits by ocean water at great depths. Large portions of the Pacific Ocean floor, for example, are pretty much free of solid carbonate sediments. Any such deposits would dissolve since the sea water there has unused capacity.

Because it must take at least some modicum of time for calcium ions and carbonate ions to combine to form calcium carbonate, the concept of supersaturation seems simple, but it is not. Supersaturation is a fairly common phenomenon not thoroughly understood in general. Some compounds can be prepared in solutions that last almost indefinitely even though concentrations are well above those dictated for precipitation of insoluble solid. Calcium carbonate ordinarily precipitates quickly in the laboratory. However, in the complex ocean environment, supersaturation to twice the normal concentration and even greater is not unusual. No completely satisfactory theory has been forthcoming although there are reasonable explanations for the effect. Nevertheless, those lie beyond our needs.

The overall picture of calcium carbonate solubility can be summarized by a few chemical equations.

$$CaCO_3 + H_2CO_3 \rightarrow Ca^{2+} \text{(aqueous)} + 2\,HCO_3^- \text{(aqueous)}$$
$$Ca^{2+} \text{(aqueous)} + 2\,HCO_3^- \text{(aqueous)} \rightarrow CaCO_3 + H_2CO_3 \text{(aqueous)}$$
$$CO_2 + H_2O \rightarrow H_2CO_3 \text{(aqueous)}$$
$$H_2CO_3 \text{(aqueous)} \rightarrow CO_2 + H_2O$$

The top reaction expresses what is occurring when limestone dissolves and when marble dissolves in rainwater for instance. The next reaction indicates how calcium carbonate precipitates in the ocean, since at seawater pH, the dominant carbon dioxide species is the bicarbonate ion as we noted in Chapter 10. At acid pH, on the other hand, most of the dissolved carbon dioxide is in the form of carbonic acid (implying CO_2(aqueous) as well) as in the top equation meaning that calcium carbonate would then dissolve. At the opposite extreme, in base, where H^+ is consumed converting H_2CO_3 to HCO_3^- the reverse situation pertains, leading to the precipitation of calcium carbonate. If carbonic acid were removed, as in the evaporation of carbon dioxide out of solution, the equilibrium would be restored through the second reaction, precipitating more solid. Increasing the pressure of carbon dioxide gas above the solution (this is the partial pressure of just the CO_2, not necessarily the total pressure) puts more carbon dioxide into solution, forming more H_2CO_3 via the third reaction, and consequently dissolving more calcium carbonate as indicated by the first reaction.

To review briefly, we noted that the solubility of calcium carbonate decreased with increasing temperatures. To this we add the fact that at warmer temperatures, there's also significantly less CO_2 in solution, compounding the direction of the change. Calcite dissolves at great depths where seawater is almost permanently cold, but at the surface, especially in warm regions, it precipitates. At great depths, pressure itself affects solubility, increasing the amount of calcium carbonate that could dissolve by as much as a factor of 2 compared to atmospheric pressure.

DISTRIBUTION

Ca^{2+} ions delivered by rivers and deep sea vents into the oceans are precipitated as calcium carbonate owing to the presence of CO_3^{2-} from dissolved CO_2 even though most dissolved carbon dioxide is present as HCO_3^-. The $CaCO_3$ is precipitated and re-dissolved several times before possibly settling "permanently." Calcium ion input is estimated to be 700 million tons (700 Mt) per year corresponding to 1.8 gigatons (1.8 Gt), ultimately of calcium carbonate appearing per year. If this were spread uniformly over the ocean floors, the deposit rate would amount to about a half gram per square centimeter every millennium. That doesn't sound like much, but considering geological times, the more appropriate way of expressing this rate is 1 to 2 m layers of solid calcite per million years. Look back at the list of geologic periods and epochs to remind yourself that many tens of millions of years are typical of times assigned.

An inventory* of carbon dioxide from the atmosphere now sequestered in the crust is as follows (the units are 10^9 tonnes or Gt):

Crustal Domain	Gt
Atmosphere	3,200
Ocean and fresh water	130,000
Living organisms and undecayed organic matter	14,500
Carbonate rocks	67,000,000
Organic carbon in sedimentary rocks	25,000,000
Coal, oil, etc.	27,000
Total	92,000,000†

If the upper part of the mantle, the layer that is presumably nearly degassed by now, is included in this tally, the total rises to about 350,000,000 Gt. The entire mantle still contains vastly more carbonate, as much as 800,000,000 Gt equivalent of carbon dioxide. In units of atmospheric totals – letting the total amount of carbon dioxide in the atmosphere now represent one unit for reference – these values can be more clearly compared as in the table here.

Crustal Domain	Relative to Atmospheric $CO_2 = 1$
Atmosphere	1
Ocean and fresh water	41
Living organisms and undecayed organic matter	4.5
Carbonate rocks	21,000
Organics in sedimentary rocks	7,800
Coal, oil, etc.	8.4

* From K. Krauskopf, *Introduction to Geochemistry*, McGraw-Hill, New York (1967).
† Additionally, methane hydrates found on some continental shelves might contain the equivalent of more than another 100,000 Gt, roughly comparable to the coal and oil total above. Methane hydrates are molecules of methane trapped within a cage of water in an ice-like structure. The methane, if released, would be slowly oxidized to CO_2 in the atmosphere.

This is what makes the carbonates special and understanding their tendency to precipitate (and re-dissolve) important. Most seashells are either the calcite or aragonite forms of calcium carbonate. Reefs are mostly carbonates.

CARBONATES IN THE OCEAN

Calcium ions can combine with the carbonate ion, CO_3^{2-} to form calcium carbonate. Calcium ions, Ca^{2+}, are one of the more plentiful chemical species in seawater, a brief listing of which is given here.

Dissolved Ions (ppm) Prevalent in Seawater and River Water (after Krauskopf)		
Dissolved Ion	Seawater*	River Water
HCO_3^-	105 ppm	42 ppm
Mg^{2+}	1,284	4
Ca^{2+}	412	15
Na^+	10,800	6
Cl^-	19,400	8

Nearly three-quarters of carbon in Earth's crust is sequestered as carbonate (predominantly $CaCO_3$) in rocks, most originating as carbon dioxide that once dominated the atmosphere. Coal, organic sediments for certain, and oil, most probably, were generated by life processes using carbon dioxide and effectively sequestered another 25% of the carbon dioxide available. Only 35-millionths of the potential carbon dioxide gas is still in the atmosphere now. If all of the carbon currently in the crust and oceans had been in an early atmosphere, that atmosphere would have been loaded with 40,000 times the current amount or upward of 16 atmospheres of carbon dioxide, enough to ensure, through the greenhouse effect, that the surface temperature of Earth remained well above the boiling point of water thus trouncing any reasonable chances of life on the planet. Is this a reasonable scenario? Recall from Chapter 2 that our neighbor Venus has an atmosphere that is 96.5% carbon dioxide with a pressure of 90 atmospheres and a surface temperature of 470°C (880°F).

There are paleontological evidence and support from ratios of isotopes that temperatures were abnormally high during the Cretaceous Period, 145–165 million years ago. Later, we will expound further about this and its relationship to how the large amounts of limestone, including chalk, among the Cretaceous sedimentary rocks leads one to infer higher carbon dioxide abundance in the past. (*Creta* is the Latin word for chalk.) But speculations about temperature variations and CO_2 fluctuations are arguably difficult to fully and confidently justify.

Much current thinking is that the primitive atmosphere, after losing its light volatiles to space, was fed carbon dioxide by degassing of Earth's interior, the mantle, and by abundant impacts from comets and meteors early in prehistory. High levels of carbon dioxide in the atmosphere would have equilibrated with earlier oceans, themselves much more acid than at present due to the formation of carbonic acid ($H_2O + CO_2 \longrightarrow H_2CO_3$). In turn, this environment would have kept calcium and magnesium carbonates soluble rather than causing their precipitation. Somehow, then – and this is still a controversial topic – life began and photosynthetic processes started consuming carbon dioxide, producing oxygen as a by-product. Such a hypothesis does explain the scarcity of carbonate sedimentary rocks dated from early Earth's history through much of the Precambrian age and into the early Paleozoic Era prior to some 570 million years ago.

---

* F. J. Millero et al. "The Composition of Standard Seawater," *Deep-Sea Research: Part I – Oceanographic Research Papers* 55, 50–72 (2008).

Expressing the relevant calcium carbonate chemistry in its least encumbered form, leaving out many complicated details, follows the fact that in the oceans, the dominant carbonate-providing species is the bicarbonate ion, HCO_3^-:

$$Ca^{2+}\left(aqueous\right) + 2\,HCO_3^-\left(aqueous\right) \rightarrow CaCO_3 + CO_2\left(aqueous\right) + H_2O$$

Often, some license is taken with this reaction to allow one to spot shifts that occur in the ocean environment, writing it instead simply as

$$Ca^{2+}\left(aqueous\right) + CO_3^{2-}\left(aqueous\right) \rightarrow CaCO_3$$

Many solids are soluble to some extent in water. The maximum equilibrium amount dissolved in a given quantity of water defines the substance's solubility in water. For most substances, warming increases the solubility. We're not talking about how rapidly a substance dissolves, but the ultimate amount that dissolves. In contrast to the common situation, the solubility of calcium carbonates – calcite and aragonite – decreases with rising temperature as we pointed out earlier. At the temperature of ocean depths, near 4°C, the solubility of calcite is about 6 milligrams per liter. In tropical water temperature, say 30°C, the solubility would be 10% lower. In the actual tropical oceans, the solubility of the solid is enhanced significantly by the concomitant drop in how much carbon dioxide dissolves in water at higher temperatures, a property common to all gases. (Think of warmed soda pop.) With removal of the (acidic) CO_2 on warming the pH rises, increasing the amount of CO_3^{2-} to combine with the calcium ions as in the second chemical equation here. Less carbon dioxide is in equilibrium with calcium carbonate at warm temperatures consistent with the presence of reefs and carbonate-laden tropical islands in warm climates but rarely where there is colder water.*

We have mentioned that as we drop below the surface of the ocean, pressure increases due to the weight of the water overhead. Every additional 32 feet of water overhead add another atmosphere of pressure, roughly 15 pounds per square inch. The solubility of calcium carbonates is also affected by pressure and, at ocean depths, can as much as double for calcite. Carbonate-containing sediments are precluded from the deep sea floor, which is under-saturated. They do accumulate at shallower depths, which are supersaturated.

Supersaturated seawater will release its CO_2. Think once more about soda pop, but in a sealed container under pressure, and what happens upon opening, that is, upon releasing the excess pressure. The liquid, for the moment, is supersaturated. Now think further about shaking that can of soda. The disturbance causes the more rapid release of the gas. This is what can happen at coral reefs. Agitating wave motion facilitates carbon dioxide loss.

The measured carbonate (ion) concentrations as a function of depth in the ocean are shown as data points in the figure for the South Atlantic (left) and the North Atlantic (right). The thin solid line sloping downward from the top edge indicates what is termed the "saturation horizon" at which depth the conditions switch from calcite (calc) being saturated or supersaturated to undersaturated. Thin dashed lines similarly describe the saturation horizon for the more soluble aragonite (arag) form of calcium carbonate. Concentrations are expressed in the figure as milligrams per liter. The data points and the short-dashed curves show, near zero depth at the tops of the graphs, concentrations higher than those allowed in a saturated solution. In the South Atlantic, the supersaturation pertains down to depths of 3 km (for aragonite) to more than 4 km (for calcite) where the data crosses the solid "saturation horizon" line. At the topmost layers, photosynthetic processes by abundant phytoplankton, for example, consume carbon dioxide leading to very high supersaturations of calcium carbonate. Deep ocean waters, below 4 km in the South Atlantic and 0.5 km in the North

---

* But there are, indeed, some live coral reefs near Alaska.

Atlantic, are undersaturated, meaning that calcium carbonates will not precipitate out. Hence, no deep water sediments.

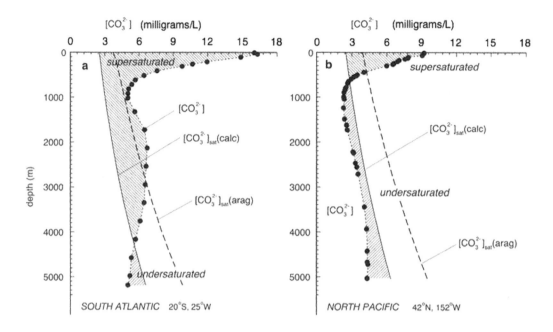

Near-surface marine algae seem to be prolific producers of carbonate. These species live, by and large, in the upper 100 m of the ocean where there is ample light. The top 15% or so of this region is called a "carbonate factory" and is where most of the production is concentrated. If shallow, warm waters are considered, algae (and also corals and mollusks) can produce sedimenting calcium carbonates in abundance. These can accumulate as undersea platforms or shelves. Not surprisingly, if the waters are not calm, if there are rough waves, frequent storms, or river deltas, the sedimentation process can be jumbled. If the waters are calm, sequential deposits can become very useful layered proxies, sources of information on local flora and fauna and ocean chemistry over eons.

Owing to the fact that insoluble carbonates will form when above certain threshold concentration levels – the saturation horizon – carbonate concentrations at depths would be much greater if not for the removal of carbon dioxide by precipitation. Acceleration was due to the evolution of efficient mineralizing species – biomineralizers – whose death would contribute to carbonate sedimentation.

CAVES

Maybe as many as 100,000 caves are to be found on our planet. Most common among these are limestone caves. One famous such cave is in Carlsbad, New Mexico, once the site of a horseshoe-shaped reef. That was apparently a quarter of a billion years ago. But then the sea above the reef evaporated or receded and the reef was covered with other deposits, slowly uplifted by geologic movements a few million years ago. Subsequently, year after year after year, rainwater, slightly acidic from CO_2 in the air, seeped into the fossil reef structure, perhaps also mixing further with trapped sulfates forming small quantities of sulfuric acid. Over the many ensuing centuries, the slow corrosive action of the acid ate away at the limestone, dissolving it, washing it away, and creating tremendous caverns. Carlsbad's "Big Room" is 400 m long, 200 m wide, and nearly 100 m high.

Seeping water laden with dissolved limestone (calcium ions and bicarbonate ions) would drip from the ceiling. As each drop slowly formed, carbon dioxide was lost and with it, as we should now understand, the ability for the limestone to remain completely in solution. Small crystals of calcium carbonate would precipitate on the ceiling. They would form conduits for the next drop, and so the deposits would build up over time forming *stalactites* and *soda straws* as seen in the pictures. The drops didn't necessarily evaporate completely. Some would fall to the cavern ground where they could similarly lose more CO_2 and deposit additional limestone, accumulating the *stalagmites* directly underneath the ceiling growths. There's one that's 19 m tall in Carlsbad.

Like the White Cliffs of Dover, these huge cave rooms were once solid with carbonates that originally had been part of the carbon dioxide rich atmosphere eons ago. However, be aware that these $CaCO_3$ deposits unlike those at Dover are not sedimented fossils.

CEMENT

Cement is the key ingredient in concrete and concrete is the most widely used building material in the world. More than a billion tons are produced per year globally corresponding to about a cubic mile of concrete.

The Romans were the first to use what would be acceptably recognized today as a cement in concrete. They combined slaked lime (calcium oxide, CaO, from heating calcium carbonate to drive off carbon dioxide) with a volcanic ash from Mt. Vesuvius. This is similar to hydraulic cement used nowadays in situations where curing under water is advantageous. The Roman structures of ages ago were remarkably durable, contributing for example to the longevity of the Coliseum in Rome.

In 1824, an English bricklayer named Joseph Aldin invented "Portland cement," obtaining a patent on his process in which limestone and clay were ground up and burned, yielding granular material that was subsequently ground again into cement. Approximately 98% of the cement produced in the United States is Portland cement. The cement, when mixed with water, gravel, sand, and frequently other materials as well, acts as the glue binding all the ingredients together.

Besides generating large quantities of carbon dioxide from burning fuel to achieve the high temperatures, around 1650°C, necessary for cement production, the limestone itself releases a roughly equal amount of CO_2 since it is slaked lime, CaO, that is the desired component in cement production. Heat decomposes the limestone's $CaCO_3$ leaving behind the CaO and generating the CO_2. In 1999, the United States generated nearly 40 Gt of carbon dioxide in the manufacture of cement and another 13 Gt for lime production. Limestone is also used in processing iron ore and in desulfurizing flue gas, producing another 8 Gt of CO_2 per year as a result.

MARBLE

Sedimentary limestone (precipitated carbon dioxide in the form of calcium carbonate), when subjected to heat and pressure undergoes chemical and physical changes. If in the proximity of hot magma or igneous rocks and buried in the crust at sufficient depths so that pressure is roughly ten times what it would be on the surface, the limestone with its impurities sometimes including fossils, changes its appearance dramatically giving rise to what we familiarly call marble. Even the "purest" of marbles, like the ones Michelangelo used from Carrara, Italy, contain significant quantities of quartz (silicon dioxide) and bits of graphite, pyrite (iron sulfide), and iron oxides. In other marbles, various silicate minerals can induce different colors when reacting with the limestone. Even fossil corals may be found. Many marbles date their "metamorphic" origin back to the Paleozoic Era as much as a half-billion years ago.

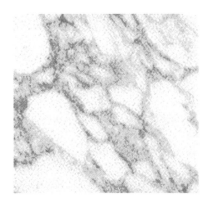

White marble

ALABASTER

Ancient alabaster was mined throughout the Middle East millennia ago. Alabaster is calcite. The term alabaster possibly originates from the ancient Egyptian. The word *a-labaste* refers to the Egyptian goddess Bast represented as a lioness and frequently depicted as such in figures on alabaster vessels. Alabaster is usually formed in limestone caves, and frequently translucent in character. A pair of striking examples of its use includes the "alabaster sphinx" at Memphis, the history of which has been lost, and the sarcophagus of Pharaoh Merenptah, the 13th and only surviving son of Ramses II, both from more than 4,000 years ago, shown here.

TRAVERTINE

When calcite precipitates from mineral springs, especially from hot mineral springs where calcium carbonate solubility is low, it can appear in the form of travertine, another natural sedimented carbonate mineral with applications in art and architecture. The ancient name for the stone was *lapis tiburtinus*, *tibur stone*, a label which was gradually corrupted to travertine. Tibur, now Tivoli, is a pre-Etruscan city near Rome. The Coliseum in Rome was constructed largely of travertine mined in Italy. The legacy of carbon dioxide can be spectacular.

Natural travertine stepped plateaus form in a process that is essentially the reverse of sink-hole and limestone cave formation. Hot mineral water seeping upward through limestone rock can deposit solid carbonates giving rise to some spectacular formations. One in Yellowstone National Park is shown in the National Park Service photo here.

EAR SAND

In our inner ears are calcium carbonate granules, colloquially sometimes called ear sand, very close to tiny sensory hairs that themselves are connected to nerves. These granules, known as *otoliths* or *otoconia*, detect motion – acceleration, orientation – and rub against the hairs sending signals to the brain. The granules, through their inertia, indicate acceleration. They are nature's invention mimicked by accelerometers in some modern electronic devices. A change in speed causes the particles to drag backward (acceleration) or forward (deceleration) against the sensory hairs. Those particles located in an up-down inner ear channel give indications of the third dimension of motion. Persons, or other animals, lacking these arrangements have troubles with balance. If the carbonate crystals become loosened from their docking sites, dizziness can result. Infants exhibit the well-known "startle reflex" when they "fall" and it is accepted that their instinctive, reflexive arm response is triggered by their otoliths' inertia. In humans, otoliths are 50–100 nanometers in size shown magnified here on the left. (That is less than the size of the period at the end of the sentence.)

 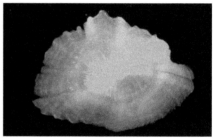

In fish, otoliths are the major receptors of sound for hearing. They also let fish know "up" from "down." Otoliths, in fish in particular, are layered in structure as is revealed in the picture here. The carbonates, as they grow, can trap trace amounts of impurities such as metals and pollutants.

Study of fish otoliths is a growing area of research. Techniques applied include isotope analysis of fossil otoliths serving as a proxy to estimate sea temperatures and other aspects of the environment in ancient times.

Fossil fish, cod-like in nature, dating from nearly 70 million years ago, have been found with otoliths somewhat in excess of 1 cm in size. The layers in these structures can be individually analyzed to track changes on a year-to-year basis in the figure shown here, exceptionally good time resolution for something that ancient. Investigators explored isotope ratios in the carbonate as proxies for temperature variations over the life cycle of the samples. One example, representative of a well-preserved collection of several specimens that likely existed within tens to hundreds of years of each other, is shown on the right.* The specimens come from South Dakota on the then coast of what is called the Western Interior Seaway that extended from the Hudson Bay area in Canada through the plains states and into the Gulf of Mexico. Oxygen (open circles) and carbon (filled circles) isotope fractionations in the $CaCO_3$ are shown on the right and left axes respectively as percent deviation from standard ^{18}O and ^{12}C abundances.

---

* S. J. Carpenter, J. M. Erickson, and F. D. Holland, Jr., "Migration of a Late Cretaceous fish," *Nature* 423, 70 (2003).

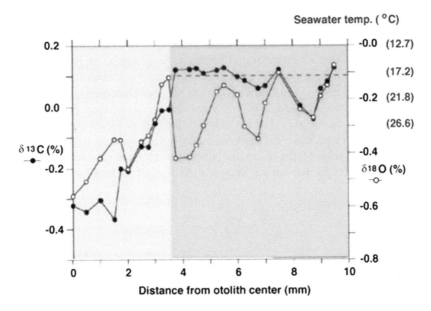

Interpretation of these data trends by the research team culminated in the following: the fish, which have a lifespan of four years, spent their first year in an estuary environment of brackish water, typified by the low enrichments of ^{18}O during which the first 3.5 mm of otolith formed. They subsequently migrated to a shallow marine environment as indicated by the larger enrichments and characterized by seasonal variations over the remaining three years of their lifetime. Without going into the detailed arguments behind the researchers' conclusions, we'll just note that the oxygen isotopic composition of the carbonate in the otoliths is representative of the aqueous environment that is nourishing the growth of calcite otoliths as the fish matured. Experiments in modern fish otoliths have shown the direct relationship between oxygen isotopic fractionation and temperature and that this relationship seems to be consistent among various species. The temperature scale on the right in parentheses is derived from the measured ^{18}O enrichments on the right axis. In contrast, the carbon isotopic fractionation (left axis) is encumbered with much more confounding contributions involving environment, diet, and metabolism, washing out the seasonal structure.

OTHER CARBONATE MINERALS

We end this chapter by pointing out that although calcium carbonates are abundant, they are far from unique. Literally hundreds of carbonate-bearing minerals are known. Among the "pure-carbonate" minerals, those not also containing halide (F^-, Cl^-, Br^-, I^-) or sulfate (SO_4^{2-}) or borate (BO_3^{3-}) or silicate (SiO_4^{2-}) for example, are many combinations of carbon dioxide with element oxides from the Periodic Table as illustrated by the underlined symbols.

H																	He
Li	Be											B	C	N	O	F	Ne
Na	Mg											Al	Si	P	S	Cl	Ar
K	Ca	Sc	Ti	V	Cr	Mn	Fe	Co	Ni	Cu	Zn	Ga	Ge	As	Se	Br	Kr
Rb	Sr	Y	Zr	Nb	Mo	Tc	Ru	Rh	Pd	Ag	Cd	In	Sn	Sb	Te	I	Xe
Cs	Ba	La*	Hf	Ta	W	Re	Os	Ir	Pt	Au	Hg	Tl	Pb	Bi	Po	At	Rn

* lanthanide group elements (15 of them).

THE ENDURING LEGACY

Carbonates are subducted deeply into Earth's mantle where they can melt, react, and often decompose to release carbon dioxide which can power volcanoes for instance. An interesting tidbit is that the subduction seems to confront a barrier at about 660 km depth associated with a seismic discontinuity.[*] The actual mechanism of this barrier effect has not been characterized but seems to be due to mantle changes in composition and structure. Somewhat shallower than this barrier is a region where carbon dioxide in the mantle can react with metallic iron or nickel present there to convert the metals to metal oxides and reduce the carbon dioxide to nearly pure carbon. At the temperatures and pressures associated with depths of several hundred meters, the released carbon can assume its higher energy crystalline form known as *diamond*. And we all are familiar with the song title…

Diamonds are forever.

VOGUE 1967

IAN FLEMING 1971

---

[*] A. R. Thomson, M. J. Walter, S. C. Kohn, and R. A. Brooker, "Slab melting as a barrier to deep carbon subduction," *Nature* 529, 76 (2016).

13 Volcanoes

Civilization exists by geologic consent, subject to change without notice.

WILL DURANT
Historian

"Volcano" is a word derived from the Roman god of fire, *Vulcan*. Volcanoes are the greatest natural source of atmospheric carbon dioxide known on Earth. The current level of atmospheric CO_2 is 400 parts per million (ppm). The upper mantle is 50–300 ppm with potentially 3–4×10^8 gigatons of CO_2 stored compared to about 3,000 Gt in the atmosphere.

OUT OF THIS EARTH

In Anatolia, Turkey, are New Stone Age paintings of volcanoes dating back 9,000 years and discovered only in the 1960s. Shown to the left of a modern painted rendition, the cave art illustrates *caldera* formation of the twin Hasan Dagi volcanoes in the town Çatalhöyük, that was occupied as far back is 11,500 years ago. *Caldera* is Spanish for cauldron, the shape of remnant structures of volcanic collapses.

A Carthaginian admiral, Hanno, sailed around West Africa around 500 B.C.E., well before the infamous Vesuvius eruption that buried Pompeii, and recorded sighting the eruption of what is now Mt. Cameroon. His West African route is shown on the map. This was the only recorded volcanic eruption on the continent of Africa for the next 2,000 years although there are over 100 volcanoes on the continent.

Despite these primitive observations, the eruption of Mt. Vesuvius in 79 C.E. is what many historians regard as the first volcanic eruption for which there was written a detailed record. That eruption buried Pompeii and nearby Herculaneum and more. There have been many eruptions of Vesuvius since and geophysical analysis indicates that there were major eruptions dating back 17,000 years, the ones in ca. 5960 B.C.E. and 3580 B.C.E. being among the largest ever in Europe. The locals near Vesuvius took their Neapolitan word for a stream caused suddenly by rain – *lava* – and used it also for the streams of molten rock pouring down the sides of Vesuvius. The word became part of Italian and subsequently, about 250 years ago, part of the English language. The picture shows the cone (*gran cono* = great cone) of Vesuvius in front of part of the remains of the Somma Volcano from 17,000 years ago. Vesuvius is east of Naples. West of Naples, by about the same distance, are the Campi Flegrei: the Flegreian Fields, home of the mythological Roman god of fire. Here was the site of an eruption 40,000 or so years ago which, powered by carbon dioxide, released 200 km³ of magma, molten rock from beneath the surface. That is enough to bury, figuratively, the entire Italian peninsula to a depth of 0.7 m. A fascinating interpretation of the Flegrei eruption is that its timing very closely matches a turning point in recent primate evolution: the explosive volcano and the probable ensuing "volcanic winter" might have been the trigger responsible for the shift in European domination from Neanderthal man to *Homo sapiens*. That speculation was proposed only very recently,[*][†] is very controversial, and no convincing studies relevant to it have appeared. The Campi Flegrei last erupted in 1538.

Worldwide, there is about one volcanic eruption per week from any one of the approximately 600 documented active sites. Estimates put 500 million people residing within 60 miles of active volcanoes. At least 1,300 volcanoes that can be documented (directly or indirectly) have erupted during the past 10,000 years. That doesn't count multiple eruptions per site. The count is problematic though. Seafloor estimates run the total much higher. Some volcanic systems have hundreds of escape cones that could be counted separately or as due to a single underlying feature. The eruption frequency has varied over Earth's history. There have been periods of high activity and also of low activity whose causes remain speculative.

Yellowstone Park is a huge plateau, mostly in northwest Wyoming, understood to have been built up by three cycles of explosive volcanic activity. These each presumably lasted only days to weeks but occurred in the past 2 million years. Melded hot ash, pumice, and other material are as much as 400 m thick in places. That's almost 25% higher than the Eiffel Tower. The ejected volume was 2,450 km³ for the first eruption which covered an area of over 15,000 km² and took place 2.1 million years ago (as determined by radioactive dating methods). The second eruption, dated at 1.3 million

[*] F. G. Fedele, B. Giaccio, R. Isaia, and G. Orsi, "Ecosystem impact of the Campania Ignimbrite eruption in late Pleistocene Europe," *Quat. Res.* 57, 420–424 (2002).

[†] W. Daies, D. White, M. Lewis, and C. Stringer, "Evaluating the transitional mosaic: Frameworks of change from Neanderthals to *Homo Sapiens* in Eastern Europe," *Quat. Sci. Rev.* 118, 211–242 (2015).

years, had a volume of 280 km^3 and the most recent eruption,* 0.64 million years ago, produced another thousand cubic kilometers of deposits. The caldera is barely visible, having been eroded over the ages, but seems to stretch over some 45 miles (75 km).

The map here shows the worldwide distribution of volcanoes. All but a few percent occur as strings of eruptions. Most are caused by subduction of plates, the slow downward plunge of one crustal layer under another with which it is colliding.

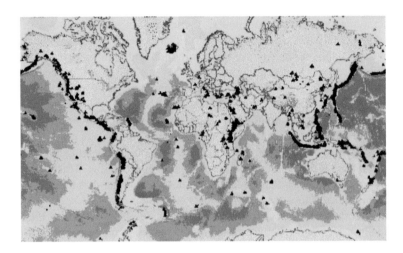

A close connection exists between volcanoes and earthquakes owing to their association with tectonic plate motions. Tectonics is the geologists' term indicating forces on and movement of Earth's crust. Here, for comparison, is a display of recent earthquake activity centers across the globe to compare with the volcano map.

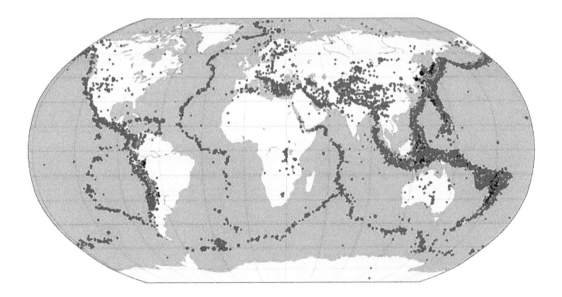

---

* There are more recent events including the "Old Faithful Geyser" but these are non-explosive.

CARBON DIOXIDE POWERED VOLCANOES

In understanding all the cumulative effects of volcanoes, the idea is to recognize that the composition of the gases from modern volcanoes is a reasonable proxy for content of primitive volcanoes. But we must keep in mind that modern eruptions recycle older sediments. It is realistic to think of the recycling as having occurred over eons except during the very early formation of the atmosphere, prior to weathering to form sediments. Throughout geological development over millions of years, huge amounts of carbonate sediments would be subducted by tectonic motion of plates.* At high temperatures, typically above 1,000°C, these carbonates would react with abundant silica (SiO_2, the analog of CO_2) and metamorphose into silicates, releasing carbon dioxide as a product. At some great depths, the presence of dissolved carbon dioxide can lower the melting temperature of minerals by hundreds of degrees compared with their melting in the absence of carbon dioxide. Carbon dioxide would be trapped in the hot syrupy blend, reducing the viscosity of the melt, easing its upward journey. Eventually, it might be released at sites of volcanoes and mid-ocean ridges. In the latter locations, crustal tectonic plates separate. As magma moves upward, the pressure to which it is subjected drops. Deeper contents can melt in the absence of any more applied heat (!). That behavior is typical of most substances that are near phase-changing conditions. A hot solid under pressure can melt if the pressure is reduced even without heating. This is a common property of material whose solid is denser than its liquid. (Water, in that regard, is unusual as ice melts when pressure is increased. Ice is less dense than – floats on – water.)

The violent and spectacular eruption of volcanoes is due in most part to the escape of gases dissolved in the lava, the escape being enabled by the reduction in pressure as the molten material approaches the surface like when a diver comes up from ocean depths. Such behavior is comparable, on a smaller scale, to the appearance of carbon dioxide bubbles when a bottle of (pressurized) champagne is opened. Roughly 5% of the weight of the magma represents dissolved gases and would result in the sudden expansion in gas volume by of a factor of 700. A common scenario is one in which the dissolved gases collect at the top of the magma until the pressure accumulation forces the gases through overlying rocks, carrying small bits of molten lava with it. Fluid, molten lava follows the opened path. Gases continue to rise to give the sometimes extended and awe-inspiring succession of explosive displays materializing at the surface. Gases can accumulate in large bubbles maybe 10 km below the surface. Long-term accumulation of gases, carbon dioxide being one of the key propellants, seems to play an important role in terms of the buoyancy and viscosity reduction of molten lava. Molten magma moves as slow as a snail's pace or to upward of a reported several hundred miles per hour at the other extreme.

For example, at Stromboli, a small island in the Mediterranean Sea north of Sicily, explosive eruptions there would last for 10 to 15 seconds and occur every 10 to 20 minutes. Such behavior is strikingly different from the steady quiescent degassing that is also occurring. Typical (remotely) measured gases from vents during quiescence were 83% water vapor, 14% carbon dioxide, plus sulfur dioxide, hydrogen chloride, and other gases. Nevertheless, explosive eruptions had more than double the carbon dioxide, 33%, with a corresponding reduction in the percentage of water vapor. Temperatures measured more than a thousand degrees centigrade matching that of molten basalt. Differences between the two types of eruptions are attributed to the origin of the explosive gases, allegedly to extreme magma depths for the violent displays compared to near-surface origin for the quiescent eruptions.†

Mt. Etna in Italy is Earth's major carbon dioxide emission source from a volcano in modern times. The annual output has been estimated at 44 Mt, a large fraction of the total volcanic output.

---

* Subduction not only fires up volcanism releasing carbon dioxide to the atmosphere, but also creates *zircons*, highly weathering-resistant crystalline residues containing traces of uranium and lead that can be used for precision dating.
† M. Vurton, P. Allard, F. Mure, and A. La Spina, "Magmatic gas composition reveals the source depth of slug-driven Strombolian explosive activity," *Science* 317, 227–230 (2007).

However, the precise values must be viewed with caution. Emission at Kilauea, Hawaii, during the extensive 2018 eruptions seems to have been in the range of 10 kilotons per day.

A few analyses* of volcanic gases in volume percent are

Gas	Kilauea, Hawaii	Momotombo, Nicaragua	Erta' Ale, Ethiopia	Erebus, Antarctica
Carbon dioxide	49%	1.4%	11%	36%
Carbon monoxide	1.5	0.01	0.4	2.3
Hydrogen	0.5	0.7	1.4	–
Sulfur dioxide	12	0.5	8.3	1.4
Hydrogen sulfide	0.04	0.2	0.7	–
Hydrogen chloride	0.08	2.9	0.4	0.7
Water vapor	37	97	77	58

David Beerling, in his *The Emerald Planet*, estimates that Hawaii's Kilauea emits 10 million tons of carbon dioxide for every cubic kilometer of basalt produced. Among other gases also found are sulfur trioxide, nitrogen, argon, and volatile salts. Sulfur trioxide, SO_3, may be from the oxidation of sulfur dioxide, SO_2. Only traces of ammonia, NH_3, are emitted implying ammonia arguably was never an appreciable atmospheric constituent. (See Chapter 2, "Early Earth.")

Although the chemicals present in volcanic gases are roughly the same among all studied volcanoes, the relative amounts vary markedly from site to site and even at the same site at different times. Much of the variation is attributed to chemical reactions that occur, reactions whose outcomes depend on temperature, pressure, and air and water encountered. The generalization is that water vapor dominates, usually amounting to more than 90% of the emissions. The carbon-containing gases are next, with carbon dioxide being the most prevalent among them. Volcanism is the greatest source of *natural* carbon dioxide supplied to the air. Approximately 150–260 million tons of carbon dioxide are released into the atmosphere each year by volcanoes.[†] This happens to be only a few percent of that emitted as the result of current human activities and less than 0.01% (0.7–1.5 ppm) of the total in the atmosphere. On the other hand, there have been vast volcanic eruptions that have continued for millions of years with a concomitant spike in the accumulation of carbon dioxide production. Keep in mind, though, that the long-term potential consumption of carbon dioxide by weathering lava greatly exceeds the CO_2 released by that given amount of lava. (See Chapter 16, "Weathering.")

Geothermal sources can be tapped to generate electricity or for heating. In 2015, nearly 13 gigawatts of geothermal electric power were extracted. That number is expected to double over the next decade. The gas emissions from the geothermal sites, aside from the steam, are almost entirely CO_2 but only 5% of what would be emitted generating the same power using coal-based plants.

Submarine geothermal activity in mid-ocean ridges could account for another 1–4 times this modern total. The estimate of all geothermal sources – volcanic, metamorphic, and others – could be as high as 300 million tons of carbon dioxide per year.

KILLER LAKES

There are three lakes in the world where carbon dioxide has accumulated in huge amounts at the bottom. These are all in Africa: Lake Nyos and Lake Monoun both in Cameroon, and Lake Kivu on the border between Congo and Rwanda. The carbon dioxide hypothetically has seeped into the lake

---

* C. Oppenheimer, T. P. Fischer, and B. Scialler, "Volcanic degasing," *Treatise on Geochemistry*, 2nd edition, Chapter 4, 111–179 (2015).

[†] *Ibid.*

bottoms from magma pools, potential volcanoes 50 miles below in the crust. The carbon dioxide is trapped there in a highly unstable situation. CO_2 dissolves in the first water it contacts, at the lake bottom. Continual feeding with no mixing of water layers leads to saturation of the bottom of the lake with carbon dioxide. The carbonated water is denser than ordinary water so it tends to remain sunken at depth. Elsewhere, as is more frequent, the carbon dioxide that seeps up goes either directly into the atmosphere or bubbles into springs that release the gas to the atmosphere. Lake Nyos is a crater lake inside a dormant volcano up in the mountains of northwest Cameroon. Its waters don't rise and fall but are quite still so that the bottom-dwelling gas is not circulated. Escape doesn't happen. The gas-loaded water is kept at the bottom by the pressure of the uncarbonated water above.

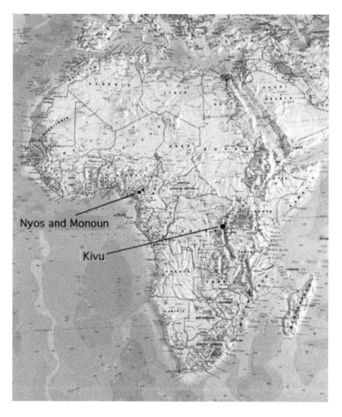

Like the candy "Mentos" placed in a soda bottle releasing the carbonation within, in August of 1986, some non-specific disruption caused Lake Nyos's carbon dioxide to erupt more than 200 feet high. A huge cloud of water and gas moving at a reported 45 miles per hour spilled down the mountain hills from the crater lake. The heavy* and suffocating gas raced into the valleys and villages below, killing 1,700 people and hundreds of animals. It is estimated that a cubic kilometer of carbon dioxide gas was released. Of more than a thousand inhabitants, only six survived. Some, living as far away as 10 km, were asphyxiated. Evidence along the lakeshore showed concomitant wave damage that reached 25 m above normal water level, with one location indicating an 80 m wave hit.

---

* Carbon dioxide is a heavy gas, 50% denser than air.

Two years earlier and about 60 miles away, Lake Monoun's carbon dioxide storage erupted killing 37 people. It was the first recorded such lake eructation ever. These phenomena could have been triggered by events as simple as a landslide, earthquake, or perhaps even a change in weather or gusty winds. As of the year 2000, Lake Monoun was 83% saturated at the greatest depths and Lake Nyos was 60% saturated with carbon dioxide at its bottom.

Tests are being run to see if large pipes might controllably release the pressure directly to the atmosphere. The first, in 2001, was done at Lake Nyos. A pipe 5.7 inches in diameter was submerged 666 feet to the deepest water. A single pipe barely keeps even with the carbon dioxide source feed. Estimates are that perhaps five pipes are needed for at least five years running in order to reach safe conditions. Results were moderately promising, keeping in mind the effort itself has the chance of triggering a burst.

In 2003, treatment of the more dangerous Lake Monoun was started. It should take years to release the estimated accumulation of 28,000 tons of carbon dioxide. Lake Nyos is more stable, but contains 500,000 tons of CO_2. Then there is Lake Kivu on the Rwanda Democratic Republic's border with Congo. Kivu holds a thousand times as much carbon dioxide as the other two lakes combined. Among explanations of mass extinctions is the possibility that the Lake Nyos scenario took place on a much vaster scale involving deep anoxic ocean water suddenly disrupted and releasing carbon dioxide.

There are now known to be nearly 500 lakes globally with appreciable carbon dioxide levels. Most of these lakes are probably well mixed, but not necessarily. A recent study* in East Java, Indonesia, revealed heterogeneous concentrations of CO_2 in a lake associated with the Kawah Ijeh volcano with occasional violent disruptive releases of gas to the atmosphere. There is so much carbon dioxide dissolved in the lake that the pH is less than 0.5, making it the most acidic in the world. A pH of 0.5 is at least 10 times stronger than "stomach acid."

FLOOD VOLCANISM AND TRAPS

In the state of Washington lies one of the youngest of a kind of layered structure known as a flood basalt. It has a volume of about 0.25 million km^3. That's 10 times the volume of the Great Lakes. In the picture here you can discern the separate layers, each being 1 to 2 m thick, that were deposited over a period of about a million years roughly 16 million years in the past. The responsible eruptions discharge vast amounts of high-temperature magma (liquid rock) non-explosively from Earth's interior. The process is aptly called "flood volcanism." Although it is non-explosive, the events occur over relatively short periods of time, a couple of million years, compared to geological eras. The annual magma outflow is high, some 1 million km^3 when active.

A number of fairly well characterized such basalt fields – called *large igneous provinces* – have appeared throughout the world during the past quarter of a billion years as indicated on the map shown here. They seem to have occurred on the average of 23 million years apart, although there is no known significance to that figure. Precise radioisotope dating is used to establish ages. One of the oldest and by far most striking of these megavolcanic landforms is in Siberia.

---

* C. Caudron, "Stratification at the Earth's largest hyperacidic lake and its consequences," *Earth Planet. Sci. Lett.* 459, 28–35 (2017).

Coffin & Eldholm, 1993

Land-Based Major Basalt "Province"	Age (My)
Columbia River (United States)	16
Ethiopia	31
North Atlantic	56
Deccan Traps (shown here, India)[‡]	66
Madagascar	88
Rajmahal (India)	116
Serra Geral (Brazil)	132
Antarctica	176
Karoo (Namibia)[‡]	183
Newark (United States)	201
Siberian Traps[‡]	252

[‡] Mass extinctions associated with this age.

Those Siberian, mostly Russian, buildups are nearly all buried under as much as 2 km of sediment deposited subsequently. Recent explorations and dating of these so-called Siberian Traps have shown the fields to be incredibly extensive, covering an area of nearly 3 million km². ("Trap" comes from the Swedish word *trapp* for "stairway.") If spread uniformly, it is enough to submerge all of Eurasia under 100–150 m of deposit.

Obviously, transport of magmatic material to the surface results in the emission to the atmosphere of large quantities of volcanic gases, of which carbon dioxide is a significant component. The association of extinctions with large igneous provinces was first recognized by Peter Vogt in 1972.* The age of the Siberian Traps, most recently (uranium-lead) dated at $251.70 \pm .04$ My, is provocatively close to that of the severe extinction "event" at the end of the Permian Period. (See Chapter 7.) This event is believed to be the largest of five extinctions,[†] one in which three-quarters or maybe even beyond 90%, of all animal species disappeared and the Triassic Period began. The traps' growth, determined from how rapidly isotopic proxies in carbonates vary, had been thought to take as long as 165,000 years, but more recent indications are that the basalt flooding was considerably more rapid: perhaps as short as 20,000 years. Such a precipitous (in geological terms) onset has further been associated with the intriguing possibility that the flood basalts, as they emerged from depths in so-called pipes, intersected coal seams causing the release of both carbon dioxide and methane in quantities 10 times larger than associated with just lava. The methane, of course, oxidizes to carbon dioxide in the oxygenated atmosphere. Higher temperatures implied by isotopic proxies are consistent with carbon dioxide levels as great as 0.003 atmospheres and affiliated ocean acidities stronger than could be tolerated by marine biota.

An alternative hypothesis is a recent controversial claim of a major meteor impact, possibly a large asteroid or comet, off the northwest Australian coast as the instrumental event. This "Bedout" impact suggests a buried crater perhaps 200 km wide.

Similarly to the Siberian Traps, a subsequent flood basalt called the Newark Traps, now accurately and precisely dated at $201.56 \pm .02$ My, ended the Triassic Period.[‡] The extent of these volcanic floods is shown as the darkened areas.

The Jurassic Period and the rise of the dinosaurs followed.

* P. R. Vogt, "Evidence for global synchronism in mantle plume convection, and possible significance for geology," *Nature* 240, 338–342 (1972).

† P. B. Wignall, "Large igneous provinces and mass extinctions," *Earth Sci. Rev.* 53, 1–33 (2001).

‡ Terrence J. Blackburn et al., "Zircon U-Pb geochronology links the end-triassic extinction with the central Atlantic magmatic province," *Science* 340, 941 (2013). The label "Newark" is associated with the trap's current remnants along the east United States coast and the west north African coast, both previously in mutual proximity as part of the super-continent Pangea.

DINOSAURUS EXTINCTUS

After dominating the terrestrial animal scene for some 185 million years, dinosaurs essentially vanished, their disappearance perpetrated in modern lore as the result of an asteroid collision with Earth. The impact centered in Mexico's Yucatan Peninsula released the element 77, *iridium*, into the environment. Iridium is related to platinum and is present at the parts per billion level making it about one-tenth as abundant as the rare platinum. But in deposits worldwide, the iridium concentration jumps up sharply and very significantly in samples dated by a variety of techniques at 66 Mya.

However, the Deccan Traps of India (see previous table) were from a major flood volcanic eruption occurring in four waves also at this time and, as with the Siberian Traps, would have released vast quantities of carbon dioxide as its roughly 1.5 million km^3 of molten basalt degassed.* In places, the flood basalts reach thicknesses of 3 km.

In the fossil record at depths corresponding to the Yucatan asteroid event some 66 million years ago, leaf stomata from ferns suggest that a "sudden" release of carbon dioxide occurred. A large CO_2 increase was the interpretation, well beyond what the Indian flood basalts' slow emanations of CO_2 implied. Explaining the proxy results invoked the rational idea that the asteroid collision caused a nearly instantaneous doubling of atmospheric carbon dioxide by melting and thermally decomposing the extensive carbonate minerals – limestone – whacked by the extraterrestrial missile. The stomata results have been challenged, though, and so remain uncertain.

So which, if either, was responsible for the extinction of dinosaurs and so on, the flood volcanoes or the asteroid impact? A very clever study has cast this ambiguity into a clearer venue.† A second even rarer platinum-like element was used to probe the timing of the Deccan Trap eruptions more precisely. The element is *osmium*, number 76. The ratio of two of its isotopes can be used as an indirect marker of origin: a proxy. That is, $^{187}Os/^{188}Os$ is approximately one-to-one worldwide on the average. Contributions from river runoff have a ratio of about 1.3 for this isotope pair. Magma, on the other hand, contains these two isotopes with a distinctly different ratio of 0.13. Volcanic traps, as we have been discussing, are mantle-derived and reflect the latter lower ratio for measured osmium isotope ratios. In studying osmium deposited in marine sediments as a function of depth (age), the contribution of runoffs from the Deccan Traps, runoffs estimated to amount to about 9% of all river feeds to the oceans, revealed a change at about the 65.5-million-year depth. The osmium trace could be measured very clearly as a function of depth along with that for the iridium. The figure here illustrates dramatically that the osmium isotope ratio began on the right, the oldest, deepest layers, with a steady decline (from a 9% ratio basalt runoff feeding the oceans) well *before* the iridium spike that appears at depths closer to the surface.

---

* Using the figure of 14 megatons CO_2 for every cubic kilometer of basalt, that is, magma, the Deccan Traps suggest that 35,000 Gt of carbon dioxide were released. That would be over 3000 ppm if staged instantaneously today. However, the duration of CO_2 release strongly influences the amount that stays in the atmosphere. Over many thousands of years of basaltic flooding, the net increase may be "only" several hundred ppm.

† G. Ravizza and B. Peucker-Ehrenbrink, "Chemostratigraphic evidence of Deccan volcanism from the marine osmium isotope record," *Science* 302, 1392–1395 (2003).

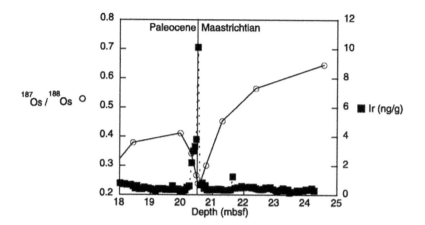

The iridium spike is allegedly from the asteroid. An interpretation, made with an improvement of ten to 100 times in the precision of radiometric dating, places the Deccan Traps' eruption(s) at about three-quarters of a million years before the iridium anomaly, causing a steady, 30% drop in the osmium isotope ratio. Coupled with these eruptions is the implied release of vast quantities of carbon dioxide into the atmosphere.

The Deccan Trap phenomenon lasted for several hundred thousand years, a duration evaluated by radiometric dating of lowest and highest samples from the thick deposits.* Recovery from the extinction of species was attained in much less time, arguing in favor of the impact explanation. Genomic DNA analysis of 48 bird species was used to date their origin in another recent study. Species as diverse as falcons, parrots, woodpeckers, pelicans, penguins, and doves, to name a few, all appeared "suddenly" within 10 to 15 million years of the asteroid impact[†] consistent with earlier fossil studies. Many species first show up as early as 1 to 3 million years after the event. Nevertheless, India's Deccan Trap degassing may have released more than twice the carbon dioxide into the atmosphere as nowadays is the total content. But the release, if spread over hundreds of thousands of years would have effected a small increase on the atmosphere's total since, on that time scale, carbon dioxide would be simultaneously removed by weathering (Chapter 16).

UNDERWATER FLOOD

The largest flood basalt domain is underwater. The Ontong Java Plateau is one of several ocean basin flood basalts. It is northeast of Australia at approximately 160° E and the equator. Its area is about that of Alaska's, five times that of the Deccan Traps, and one-third larger than the Siberian Traps. The thickness of the deposit averages 30 km and its volume is a mind boggling 50 million km³. It dates back to about 120 Mya with indications that there is also a younger component 90 million years old. That volume signifies a release of 0.5 million Gt of carbon dioxide, more than 30 times the Deccan release. The origin of this large igneous province remains unsettled. An extraterrestrial source was ruled out by a study of the platinum family element abundance as done for the Deccan Traps but no anomalies were found. Worth noting is that such large volcanic plateaus are also present on the moon, Mars, and Venus.

---

* B, Schoene et al., "U-Pb geochronology of the Deccan Traps and relation to the end-cretaceous mass extinction," *Science* 347, 182–184 (2015).
† E. D. Jarvis et al., "Whole-genome analyses resolve early branches in the tree of life of modern birds," *Science* 346, 1320 (2014).

EARTHQUAKES

In California is a large remnant volcanic caldera known as the Long Valley Caldera. In its southwest corner is Mammoth Mountain, a ski resort about 60 miles northeast of Fresno. The site is volcanically active and has been for approximately 4 million years. Except it hasn't erupted in nearly two centuries. However, underneath the mountain there is a lot of minor earthquake activity, perhaps a couple of hundred events per month at times. In the early 1990s, forest rangers noted that there were areas of dead (and dying) trees at the foot of the mountain. Studies eliminated both drought and insect damage as causes. It was then found that the soil-trapped air contained between 20% and 95% carbon dioxide compared to 1% in regions away from the dead tree affliction. That's enough to kill the trees, essentially by asphyxiating them at the roots where oxygen is needed. Approximately 100 acres of plant life have been destroyed by releases of perhaps 1,200 tons of carbon dioxide per day. A decade earlier, that amount was three to four times greater. The carbon dioxide very likely comes from 5–9 km deep magma intrusions into limestone-rich sedimentary rocks underlying the region.

A more dramatic display of carbon dioxide power was a series of nine earthquakes of magnitudes between 5 and 6 over about a month period in northern Italy in 1997 and again in 2016. Interspersed with the major quakes were thousands of aftershocks. Geophysicists confidently believe that this sequence was driven by the seismic release of pent-up carbon dioxide, known to have been trapped several kilometers below the Apennine mountain surface.* The CO_2 was sealed at these depths by an overlay of non-porous compressed layers of minerals at pressures perhaps as high as 700 atmospheres. That value corresponds to five tons per square inch. Seismic activity, perhaps a slight slip along the overlying faults, is assumed to have opened the seal by providing porous escape access.

---

* S. A. Miller et al., "Aftershocks driven by a high-pressure CO_2 source at depth," *Nature* 427, 724–727 (2004).

14 Photosynthesis

It's not easy being green.

KERMIT THE FROG

SIGNIFICANCE

Oxygen is used by living things to convert carbohydrates and fuels into carbon dioxide, water, and energy. If it weren't for the oxygen produced by photosynthesis, the atmosphere would be unsustaining in about 3,000 years. And although microscopic phytoplankton account for 1% of marine biomass, they are responsible for nearly half the photosynthetic activity on our planet.

Photosynthesis, as its name suggests, refers to any natural or artificial process of forming material that involves the use of light. Common usage and tradition apply the term more specifically to the use of light by green plants in the manufacture of carbohydrates and other chemical compounds within cells. Plants employing photosynthesis range in dimensions that range over a factor of 100 million, from the smallest phytoplankton (less than 1 micrometer; more than 25,000 to the inch) to giant trees such as redwoods (including 100 m tall sequoias). It has been estimated that 440 Gt of carbon dioxide per year are consumed by photosynthesis in terrestrial plants. Marine plants consume 370 Gt more. To put these magnitudes in perspective, remember that the total atmosphere content of carbon dioxide is 3,200 Gt. Not all the consumed carbon dioxide is stored though. Only about half stays put, the rest soon returns to the surroundings due to plant "respiration" (discussed in more detail in the next chapter).

It hasn't always been like this. But photosynthesis has been the lead actor in the multibillion-year-old carbon dioxide legacy. Photosynthesis has been referred to as the key shaper of evolution. As a vital process, it dates back more than 2 billion years before the present, to the "Archaean" Eon. At about that time, oxygen-evolving *cyanobacteria* were thriving, consuming carbon dioxide, forming matted conglomerates, leaving them behind as fossil *stromatolites*, a picture of which is shown here. Limestone is often part of stromatolite composition but whether the origin is inorganic or biological is debated.

An astounding environmental outcome of this development was that the rising oxygen concentrations from (alleged) stromatolite metabolism led to oxidization of the abundant greenhouse gas, methane, CH_4, also present in the primitive atmosphere. The loss of greenhouse gas arguably led to severe drops in temperature, averaging at its "worst" perhaps 50 degrees below zero Centigrade. There is growing evidence, discussed in Chapter 17, that subsequent cycles – ice ages interspersed with extreme warming periods – occurred a number of times, most recently during the Precambrian Era about 750–590 Mya. The transition from warm climate to "snowball Earth" might have taken as little as a thousand years to evolve, freezing oceans and extending glaciers even to equatorial regions, eliminating many lifeforms during the freeze-out cycle. Carbon dioxide's ups and downs have been profoundly influential.

DISCOVERY

Serious study of plant growth dates back a few hundred years. Earlier it was mentioned that Jan Baptista van Helmont planted a willow tree in a bucket and watered it using only rainwater for five years. After five years, the tree had grown significantly, gaining 164 pounds, but the bucket lost negligible weight, several ounces. Van Helmont concluded that the tree growth must have come from the water fed to the tree. Stephen Hales, the English botanist, observed that it was "air" that the plants used for growth. In the 1770s, the English chemist Joseph Priestley did some experiments with "fixed air." After burning a candle until the combustion self-extinguished, he showed that a mouse, placed in the atmosphere of fixed air, died. However, a mint herb continued to thrive in the fixed air for weeks. Subsequent to the green plant's growth, Priestly found that a candle could burn again and a mouse could now breathe successfully in the replenished air. Toward the end of that decade, the Dutch physician Jan Ingenhousz was inspired by Priestley's experiments. Ingenhousz himself performed plant growth experiments in 1779 showing that sunlight was necessary and that only the green parts of plants participated in replenishing the air giving off oxygen. He also demonstrated that, *in the dark*, plants, like animals, consumed oxygen and gave off carbon dioxide. He wrote his results that year in a book quaintly entitled *Experiments upon Vegetables, Discovering Their Great Power of Purifying the Common Air in Sunshine and Injuring It in the Shade and at Night.*

Jan Ingenhousz

In 1783, a Swiss minister named Jean Senebier observed that plants utilized carbon dioxide that was "dissolved in water." After the turn of the century, a Swiss chemist, Nicolas de Saussure, the son of an agricultural scientist, concluded from quantitative weight gain studies on the organic matter and oxygen produced by plants that they required not only carbon dioxide, but water as well.

Pierre Joseph Pelletier and Joseph Caventou

In 1817, two French chemists, Pierre Joseph Pelletier and Joseph Caventou, both sons of pharmacists, used alcohol as a solvent to separate out the green component of plant leaves responsible for plant respiration. They named the substance *chlorophyll* from the Greek for green (chloro) and for leaf (phyllon). (A few years later they made their most important discovery, *quinine*, from the bark of the cinchona tree.) In the middle of the nineteenth century, Julius Mayer, a German physician, represented photosynthesis as

$$CO_2 + H_2O + light \rightarrow O_2 + organic\ matter + chemical\ energy$$

Julius Mayer

Jean Baptiste Boussingault

In 1864, the French agricultural chemist, Jean Baptiste Boussingault, found out that the ratio of oxygen generated to carbon dioxide consumed in photosynthesis was essentially one-to-one. That year, the German botanist Julius von Sachs did an elegant experiment using the ability of iodine to detect starch. Starch is mostly a sugar polymer, a chain of sugar molecules. When iodine comes in contact with starch, the starch turns characteristically purple. Sachs kept some green leaves in the dark for many hours so that their metabolism consumed any starch that might be in the leaves. He then covered half of each leaf and exposed the other half to sunlight. After a while, he exposed the entire leaf to iodine and only the exposed portions turned color. What did this demonstrate? Light is needed for photosynthesis; here, the production of starch. In 1865, Sachs proved that chlorophyll was present only in certain cellular bodies, later called *chloroplasts.*

Julius von Sachs   Chloroplasts   Melvin Calvin

MECHANISM OF THE SYNTHESIS

Early ideas of how photosynthesis worked were quite simple. In the first quarter of the twentieth century, it was thought that light split carbon dioxide into oxygen and carbon, the latter subsequently reacting with water to produce organic compounds including carbohydrates (like sugars and starches). However, to confound the apparent simplicity of such an idea, some bacteria were found which could use carbon dioxide and water to synthesize essentially the same organic compounds that were made by photosynthesis, but the bacteria didn't need light. In 1931, the Dutch microbiologist Cornelis van Niel found bacteria that could utilize the carbon dioxide in the presence of light, but that did not evolve oxygen. The hydrogen needed was extracted by the bacteria from H_2S, hydrogen sulfide. The similarity between sulfur and oxygen – their relationship in the Periodic Table – induced van Niel to suggest that the oxygen produced by photosynthesis ordinarily came from H_2O, not carbon dioxide.

In 1941, oxygen enriched in the rare isotope ^{18}O became available. In Chapter 1, we noted that ordinary oxygen is composed of only 0.2% of this isotope. Availability of ^{18}O enabled Americans Sam Ruben and Martin Kamen to use water enriched in the heavy oxygen isotope as a tag for following photosynthesis reactions. The oxygen observed to be evolved as gas had the same percent composition as the tag used in the water, but did not have the composition of the carbon dioxide's oxygen. This showed conclusively that the oxygen released was not from the carbon dioxide, a finding that was very surprising. Ruben and Martin also tried to trace the path of carbon, but the only viable method was to use a very short-lived (20 minutes half-life) radioisotope of carbon, ^{11}C, making the experiments very difficult. Furthermore, the carbon ended up in a variety of organic compounds that could not be separated from each other sufficiently rapidly. In a few years, Ruben and Kamen discovered the long-lived radioactive isotope (half-life 5,700 years) of carbon, ^{14}C, radiocarbon (Chapter 5). It enabled Melvin Calvin, an American biochemist, to begin tracing the steps of photosynthesis by which carbon dioxide and water, under the empowerment of light, are able to assemble large complex organic matter such as sugars and starches and eventually cellulose and lignin, a quest launched nearly 300 years earlier by van Helmont.

CALVIN PHOTOSYNTHESIS CYCLE

By 1945, the availability of radiocarbon (^{14}C) in conjunction with simple techniques for separating complex mixtures into their chemical components enabled American chemist Melvin Calvin and his colleagues to begin working out the detailed path of carbon in photosynthesis.

A simplified description of the technique employed goes as follows. Green algae are exposed to light for a very brief time in the presence of carbon dioxide that has some radioactive carbon dioxide (symbolized as *CO_2)* mixed in. The algae are then rapidly dumped into methanol (wood alcohol) which quickly quenches all steps of the photosynthetic process. The various chemical components are next separated from each other and scanned to see where the radioactive carbon appears. If exposures to light and tagged *CO_2 were even as short as one-and-a-half minutes, the ^{14}C could appear in more than a dozen different chemical compounds.

Logically, if the reaction is stopped soon enough, only the first chemical or chemicals produced will contain radioactive carbon. Nevertheless, even after just five seconds of photosynthesis, the radioactive carbon shows up in a number of different chemicals. Prominent among these is one called *phosphoglyceric acid* or PGA. Its chemical structure is indicated on the right here in a diagram that represents what is now known about how the radioactive carbon dioxide, indicated as *CO_2, is first incorporated into the more complex structure. By reacting with the 5-carbon compound[†] "RuBP," a carbon dioxide goes on to produce two PGAs (only one of which, of course, will be radioactively tagged). Isotopic fractionation for this early step ranges between 2.5% and 2.9% enrichment[‡] of the lighter isotope (depletion of ^{13}C). An enzyme,[§] a natural catalyst protein called "rubisco," speeds this up. (Rubisco is also involved in photorespiration and in metabolism, see Chapter 15, where its detailed molecular structure is shown as well. The protein can constitute as much as half of the protein in leaves.)

$$
\text{*CO}_2 + \text{H}_2\text{O} + \begin{array}{l} \text{CH}_2\text{OPO}_3^{2-} \\ | \\ \text{C=O} \\ | \\ \text{CHOH} \\ | \\ \text{CHOH} \\ | \\ \text{CH}_2\text{OPO}_3^{2-} \end{array} \xrightarrow[\text{"rubisco"}]{\text{enzyme}} \begin{array}{l} \text{CH}_2\text{OPO}_3^{2-} \\ | \\ \text{CHOH} \\ | \\ \text{*COOH} \end{array} + \begin{array}{l} \text{CH}_2\text{OPO}_3^{2-} \\ | \\ \text{CHOH} \\ | \\ \text{COOH} \end{array}
$$

RuBP   PGA   PGA

The outcome was far from what was anticipated. A more straightforward scheme would have had CO_2 somehow picking up hydrogens from water to form a 1-carbon species, something like CH_2O, and then doubling and tripling, for example, to perhaps form $[CH_2O]_2$ and $[CH_2O]_3$. Instead, it was found that CO_2 combined with a 5-carbon molecule already present in chloroplasts to form a pair of 3-carbon molecules. No oxygen was released. The elaborate cycle that Calvin revealed with ingenious experiments showed that three carbon dioxide molecules and a number of photons of light were required, not only to generate another 3-carbon sugar, *triose P*, but also to regenerate the initial 5-carbon "RuBP" and a number of other compounds needed to supply the energy requirements to assemble that triose P. These other compounds are abbreviated as NADPH and

---

* A radioactive atom within a molecule is conventionally flagged by an asterisk.
† Ribulose 1,5-biphosphate.
‡ D. L. Royer, R. A. Berner, and D. J. Beerling, "Phanerozoic atmospheric O_2 change: Evaluating geochemical and paleo-biological approaches," *Earth Sci. Rev.* 54, 349–392 (2001).
§ Enzymes can be likened to those functionaries that can perform quick marriages or divorces. Their role is to bring two others together to make a couple or divorce a couple without being consumed themselves. The twosome can be pushed in either direction. Likewise, an enzyme facilitates a combination or de-combination of parts without itself being used up.

ATP.* What follows is a simplified and abbreviated version of the photosynthetic pathways which are much more fascinating and intricate when viewed in detail. The American biochemist and prolific author Isaac Asimov published a comfortably manageable book *How Did We Find Out about Photosynthesis?* that can be consulted for additional reading. A much more extensive exposition is *Energy, Plants and Man* by David Walker. Deeper scientific insight may be found in Robert Blankenship's *Molecular Mechanisms of Photosynthesis*.

During the full Calvin cycle to synthesize a 6-carbon sugar, a total of 12 NADPH is consumed (forming $NADP^+$) and 18 ATP are consumed (forming ADP^+ and the phosphate ion, abbreviated as P^-). They require a variety of enzymes, nature's catalysts, to work their magic.

$$6CO_2 + 18ATP + 12NADPH + 12H_2O \rightarrow C_6H_{12}O_6 + 18ADP + 18P^- + 12NADP^+ + 6H^+$$

Let us observe some implications in this equation before continuing with carbon dioxide's metamorphosis. For example, this scheme now introduces the need for consideration of "phosphate" as a requirement. Also, where is the oxygen by-product? For this chemical equation to represent a cycle, the ingredients on the left of the arrow, those needed to assemble the carbon dioxides appropriately, have to be resupplied, that is, regenerated as was the case mentioned for the RuBP. In one of the accompanying cycles, the necessary ATP can be reassembled from the products ADP, P^-, and H^+ given the right enzymes and the needed energy. We can write this as

$$ADP + P^- + H^+ \rightarrow ATP + H_2O$$

To restore the NADPH from $NADP^+$, a second ancillary cycle is the one that generates molecular oxygen in a manner abbreviated as

$$2NADP^+ + 2H_2O \rightarrow 2NADPH + 2H^+ + \mathbf{O_2}$$

Both of these last reactions use light as their energy source. O_2 comes from the water! In fact, the only role that photosynthesis has in the generation of oxygen is that it provides a long and very intricate path for eight photons of light to be absorbed, their energy conducted to a reactive site where two waters held in close proximity pick up electrons, rearrange, and release O_2.

Putting these last three reaction equations together to get the net change gives us the simplified result

$$6CO_2 + 6H_2O \rightarrow C_6H_{12}O_6 + 6O_2$$

Returning to the discussion involving the CO_2 itself, a very much abbreviated version of what goes on is diagrammed here. To keep relatively simple what is extraordinarily intricate, only the number of carbons in a participating species is indicated, except for carbon dioxide itself. There is actually a variety of different important intermediates lumped together this way. C_3 could be any of several different 3-carbon compounds. The process has been divided into three diagram regions which could be abridged by carbon-counting arithmetic labels $1+5=3+3$, $3+3=6$, and $6+3\times3=3\times5$.

---

* Nicotinamide (NADPH) and adenosine triphosphate (ATP).
† Adenosine diphosphate (ADP).

$$CO_2 + C_5 \longrightarrow \boxed{C_3} + C_3$$
$$CO_2 + C_5 \longrightarrow \boxed{C_3} + C_3$$
$$CO_2 + C_5 \longrightarrow \boxed{C_3} + C_3$$
$$CO_2 + C_5 \longrightarrow \boxed{C_3} + C_3$$
$$CO_2 + C_5 \longrightarrow \boxed{C_3} + C_3$$
$$CO_2 + C_5 \longrightarrow \boxed{C_3} + C_3$$

$$\boxed{\begin{matrix}C_3\\C_3\end{matrix}} \longrightarrow C_6 \text{ "a six-C sugar"}$$

$$\boxed{\begin{matrix}C_3\\C_3\end{matrix}} \longrightarrow C_6 \qquad C_6 + \begin{cases}C_3\\C_3\\C_3\end{cases} \longrightarrow \begin{matrix}C_5\\C_5\\C_5\end{matrix}$$

$$\boxed{\begin{matrix}C_3\\C_3\end{matrix}} \longrightarrow C_6 \qquad C_6 + \begin{cases}C_3\\C_3\\C_3\end{cases} \longrightarrow \begin{matrix}C_5\\C_5\\C_5\end{matrix}$$

On the top left in the figure is the overall need for six carbon dioxides, each of which individually attaches to the 5-carbon species (RuBP) in the plant followed by the break-up of each combination into a pair of 3-carbon compounds as we saw earlier. These combinations do not occur simultaneously but are lumped together as a way of compressing the details of the overall process. Half of the C_3 species (boxed) are eventually paired up to give three 6-carbon sugars, abbreviated here as C_6. We also have the other half of the C_3 species (circled) remaining on the diagram. Why do we need all these extra pieces if we've made three sugars from six carbon dioxides? Because if we ever want to perform another cycle of sugar production, we need six more C_5 (RuBP) species as already mentioned. Where are these to come from? As indicated in the bottom right half of the illustration, two of the three 6-carbon sugars and the six "extra" 3-carbon species can react (also in a quite complicated manner involving a number of steps) to re-form the six C_5 species, RuBP, needed for the next round of sugar production with new CO_2. Only one net sugar has been produced. Left out of this simplified diagram is the remainder of the cycle involving the necessary water and the fact that oxygen is produced from it as a by-product. We can now look at the actual steps in a bit more detail to explore the fascinating complexity of the photosynthetic assembly line.

The common 6-carbon sugar is *glucose* whose molecule's ring structure is shown here.

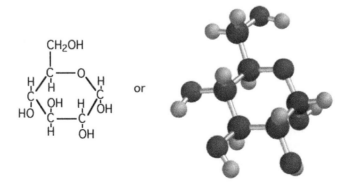

The overall process to this point can be summarized, leaving out a large number of interme-diate details. Within the plant leaf chloroplast, the 3-carbon chemicals are the building blocks for sugars and starches and more. When transported out of the chloroplast, the 3-carbon sugars pair up to begin the sequence of steps that leads to formation of the 6-carbon sugar glucose and subsequently to a 12-carbon sugar *sucrose*. But only one out of six of the 3-carbon molecules made is used for these syntheses. The other five are needed to regenerate the RuBP for the cycle to continue another loop.

Early in the complete Calvin cycle scheme, where the 3-carbon intermediates appear, there is a need for about one thousand water molecules per carbon dioxide consumed. Yes, one thousand. This is not a chemical need, but provides a molecular environment enabling the reaction to proceed efficiently. A wet neighborhood is necessary at the molecular level. A dry one is trouble for the Calvin cycle.

MORE WATER, LESS WATER, ANY ALTERNATIVE PATHS?

Hugo Kortshak, a physical chemist and plant physiologist, working as an Associate Biochemist for the Experiment Station at the Hawaiian Sugar Planters' Association, decided to use Calvin's 1945 experimental approach to study photosynthesis in sugarcane. Kortshak, with co-workers, published their studies on radiocarbon fixation in sugarcane in 1954. Much to their surprise, the first radio-actively labeled products to appear were not PGAs, but the two C_4 species, *malate* and *aspartate*. This led a number of research groups to unravel a variation on the Calvin scheme which we will explore briefly.

Where water is scarce or even just *when* water is scarce, photosynthesis becomes more difficult to achieve by the Calvin C_3 process. But there are plants that have evolved an alternative photosyn-thetic path, one requiring half the water of the 3-carbon sequence. It is a "carboxylation–decar-boxylation–*transport*" sequence that gives a boost to photosynthesis. In this alternative scheme,

an enzyme combines carbon dioxide with *phosphoenolpyruvate* (hence, it is called a *carboxylation* reaction) to form *oxaloacetate,* a 4-carbon intermediate. The C_4 intermediate is then usually con-verted to another C_4 species, *malate,* which is effectively delivered to chloroplasts. Here, interest-ingly, CO_2 is detached from the malate and dropped back into the 3-carbon Calvin cycle as before. The chloroplasts do not allow the carbon dioxide to exit and consequently, the carbon dioxide con-centration can slowly build up to provide fuel for the C_3 cycle already detailed. The remnant from unlinking CO_2 (the *decarboxylation* reaction) from malate is *pyruvate*. Pyruvate returns for another round after a phosphate is added making it again phosphoenolpyruvate on the left side above. C_4 plants besides sugarcane include maize and some tropical grasses.

The 3-carbon path does not become more productive in intense light and does best at temperatures of 15–25°C. In contrast, the 4-carbon alternative does *not* saturate and does best at 30–45°C. An obvious interpretation is that the C_4 path is adapted to very warm climates and aridity. Forty-five%

of terrestrial Earth is dry land; almost 60% of Australia is. Of course, some water is necessary. A 2011 study suggested that the La Niña weather phenomenon stimulated sufficient rainfall over arid regions of the southern hemisphere that growth of C_4 vegetation measurably increased carbon dioxide drawdown* by a significant percentage compared to recent years. But less obvious is that it is also adapted to be effective in low CO_2 levels as might occur in aqueous, subsurface environments. Marine microbes that cannot compete for carbon dioxide for photosynthesis near the ocean's surface are physiologically constructed to take up and store the compound at night when there is no competition. In the daytime, they need only the sunlight to process the CO_2 that has been accumulated in close proximity to the rubisco within the cell. Still another variation that has evolved, one to minimize the loss of water, involves absorbing CO_2 at night, and converting it to oxaloacetate and malate just discussed. During daylight, plant leaf stomata close and sunlight enables the subsequent decarboxylation and thereby photosynthesis. Aloe and pineapples are examples of plants that use this variation. The existence of some such sort of process was first suspected by chemist Nicolas de Saussure all the way back in 1804.

Arguments are that the C_4 front-end gorging – the alternative to the direct C_3 process – developed when land plants evolved. Prior to this, the atmosphere was rich in carbon dioxide and oxygen was much rarer. The photosynthesis *rubisco* enzyme that facilitates the very first carboxylation step (addition of CO_2 to the 5-carbon "RuBP") operates at maximum efficiency in such an environment. If the atmospheric composition is reversed – rare carbon dioxide and plentiful oxygen – the enzyme's activity would be compromised by aggressive competition from the oxygen. Under such otherwise unfavorable circumstances, a different enzyme[†] which could carboxylate (add carbon dioxide) *phosphoenolpyruvate*, even at high O_2 levels, serves to launch the carboxylation–decarboxylation–transport sequence that delivers CO_2 to the chloroplasts of plant leaves circumventing the adverse environment.

The bottom line: C_4 plants are capable of concentrating CO_2 and consequently are less influenced by changes in atmospheric concentrations. There is some evidence that C_4 plants were favored in the last ice age. That is, they flourished, relatively speaking.

For a thorough review of compensation mechanisms, see "CO_2 Concentrating Mechanisms in Algae"[‡] and "Advances in understanding the cyanobacteria's CO_2-concentrating mechanism (CCM)."[§] Plant photosynthesis types serve as environmental proxies. Finding fossil species of one vegetation type in long-buried sediment layers above and/or below fossil species of the other type is an indication of how the local environment may have changed over time. These differences, between photosynthetic pathways for different species, become a tool for studying temperature changes in the distant past.

LIGHT

Many, many steps and intermediates are involved in the detailed photosynthetic process. And there are a number of further variations on the main theme, in addition to the C_3 and C_4 cycles mentioned previously. Light is absorbed by *chlorophyll*, a molecule which is extremely efficient for capturing "visible" light from the sun reaching Earth. Chlorophyll acts like an antenna. The chlorophyll molecule

---

* B. Poulter et al., "Contribution of semi-arid ecosystems to interannual variability of the global carbon cycle," *Nature* 509, 600 (2014).
† Phosphoenolpyruvate carboxylase.
‡ M. Giordano, J. Beardall, and J. A. Raven, *Ann. Rev. Plant Biol.* 56, 99–111 (2005).
§ G. D. Price, M. R. Badger, F. J. Woodger, and B. M. Long, *J. Exp. Biol.* 59, 1441–1461 (2008).

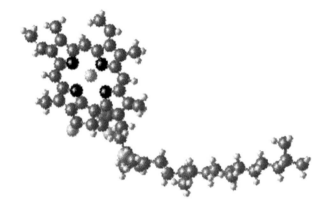

has a magnesium ion, Mg^{2+}, at its core. There are a few kinds of chlorophyll. One absorbs both red light (wavelengths at around 660 nanometers) and blue-violet light (at ~430 nanometers). Since red and blue are taken from "white" sunlight, and white sunlight is the fully mixed set of colors associated with the rainbow's spectrum, what's left over appears green to our eyes. There are actually two slightly different chlorophylls involved in most photosynthetic setups. Their makeup differs slightly, but it's enough to shift the colors absorbed. The three-dimensional rendition of chlorophyll looks like the figure here.

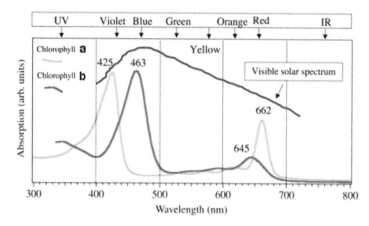

The graph here shows the intensity of sunlight over the various wavelengths that we're familiar with. Also shown are the efficiencies for two chlorophylls, a and b, in absorbing the colors.

Today, there are also known to be further variants called chlorophyll c_1, chlorophyll c_2, chlorophyll d, and chlorophyll f. There are also red algae and cyanobacteria (blue-green algae) that use different "antennae" than chlorophylls. They use simpler light-absorbing species, among the most common of which are conglomerations of *phycoerythrin,* whose structure is somewhat similar to that of a tailless chlorophyll as the structural diagram here tries to convey.

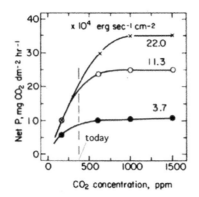

Among the many steps whose details we will not visit in our photosynthesis chemistry discussion is the regeneration of NADPH from the action of light on product NADP in the presence of water, generating oxygen from the water. Recall, oxygen does not come from the carbon dioxide. Oxygen comes from the water.

Our Kermit the Frog quote at the beginning of this chapter is succinctly poignant about photosynthesis: It's not easy being green.

CHANGES

The rate of photosynthesis depends, of course, on how much carbon dioxide is present in the atmosphere. After all, if there were no CO_2 the rate of photosynthesis would be zero. We are now at 400 ppm CO_2. The dependence of the reaction rate on concentration varies from species to species. For example, in tobacco, the net rate hits zero (photosynthetic consumption of CO_2 equals release of CO_2 by respiration) at 50 ppm carbon dioxide; maize (which uses the C_4 chemistry of the photosynthesis cycle adapted to low concentrations of carbon dioxide) can continue production down to a few ppm CO_2. Geologic records imply that the carbon dioxide level has not been anywhere near that low during the past 400 million years. For proxy illustration, finding remnants of maize buried in layers that date back to 400 million years would suggest that the CO_2 level (necessary to grow maize) did not drop below a few ppm. However, interpretation of such hypothetical records also depends on the physiology of plant photosynthesis for ancient plant life, and to what extent modern maize represents ancient maize behavior.

As far as increasing carbon dioxide concentrations go, the rate of photosynthetic production *via* consumption of carbon dioxide grows with increasing levels. However, the rate seems to saturate, to stop increasing, at about double the current level. Production, P, from spinach plants is shown here for three different levels of light (expressed as energy [ergs] incident on a unit area, cm^2, a combination which is the same as a milliwatt per square meter, approximately the amount of energy that a 100 watt light bulb shines on a square meter area located 100 m away from the bulb).

At still higher CO_2 levels, the photosynthesis rate begins to fall; for some plants this sets in at 0.02 atmospheres (20,000 ppm or 2%) carbon dioxide. Yet algal growth seems to be able to be sustained up to about 30% carbon dioxide. To appreciate this value, be reminded that the current level of oxygen in the atmosphere is 20% at sea level.

HIDDEN CARBON DIOXIDE

Nearly 90% of *terrestrial biomass* carbon, some 560 billion tons, is stored in trees. It *was* all CO_2.

Plant leaves during daylight photosynthesis accumulate *starch*. Starch – essentially a helix-like chain of C_6 glucoses – is used as a nutritional storage substance in vegetation. But, if instead of linking the glucoses in the manner that gives starch, we instead were to link glucoses slightly differently to form a linear chain as shown here, we get *cellulose*. Cellulose is the most abundant organic compound in the biosphere by weight, containing more than half of all organic carbon originating from CO_2. Cellulose's chain molecular structure gives high tensile strength to the resulting fibers. Comparable sections of the chains (or polymers) of glucose in starch and glucose in cellulose illustrate the different linkages in the figure. The three-dimensional helical structure in starch (not illustrated) is important in making the glucose molecular building units accessible for use as "food." In additional to starch and cellulose, yeasts and bacteria store glucose in yet another polymeric structural family called *dextrans* which are branched rather than linear chains of glucose.

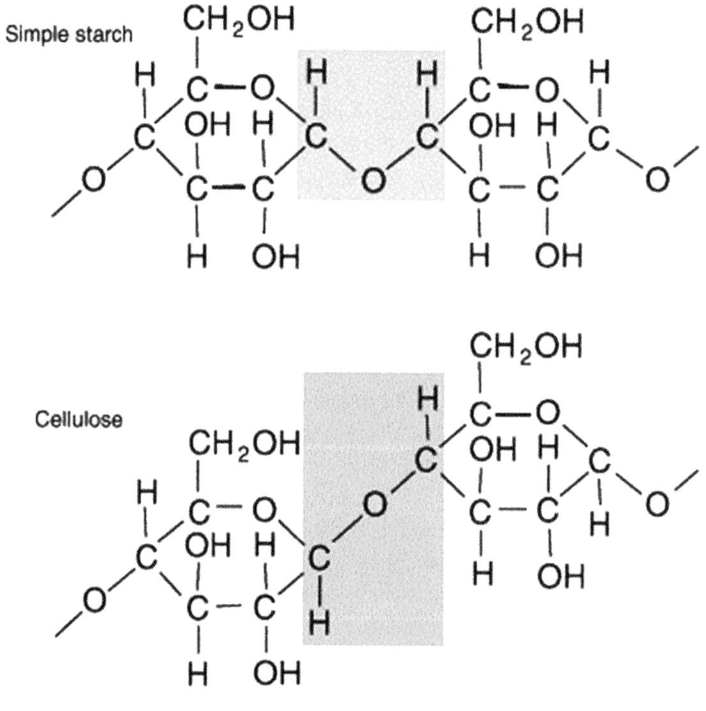

The second most abundant organic substance on the planet's land surface, after cellulose, is the structural component of wood: *lignin*. Lignin is not a single structure but a category of quite complex assemblages. It provides plants with their ability to persist in a terrestrial environment where gravity, wind, and weather would otherwise prove to be very disadvantageous to survival. Lignins (and related *lignans*) account for 20–25% of the weight of a woody plant and 30–40% of the organic carbon contained therein. One picture of what a section of the lignin polymer looks like with its notable complexity is illustrated here just for thoroughness. The details are unimportant to us.

The reason to even mention lignin is that its origin is another legacy of carbon dioxide. The very first step in the biochemistry of a plant making lignin involves the combination of two simple phosphate derivatives of sugar-like chemicals. They are *erythrose-4-phosphate* and *3-phosphoglycerate*.

erythrose-4-phosphate   3-phosphoglycerate

These are both made from glucose (or starch) by plants through a series of many steps after which even more complicated chemistry begins to assemble the huge lignin molecule, of which a section was just shown. Bear in mind that the glucose from which this starts is all assembled by photosynthesis with carbon dioxide as the feed material.

Ocean plant life is a small fraction of plant population now; maybe one-third of 1%. But the typical turnover is so rapid that their contribution to photosynthesis is comparable to that of all terrestrial plant life. In considering marine plants, though, carbon dioxide's fate becomes complicated because plant respiration (see the next chapter) in the oceans restores most (90%) of the CO_2 to surface waters for re-use by phytoplankton. CO_2 is also returned to the atmosphere directly. Nonetheless, the rest does sink as a nearly continuous drizzle of microscopic sedimentary particles amounting to removal of about 100 million tons of carbon dioxide equivalent per year.

NIGHT AND DAY, SUMMER AND WINTER, UP AND DOWN

As photosynthesis waxes and wanes with sunlight hours and seasons in a forest, the carbon dioxide concentration can fluctuate as well. It has been found to reach higher than 400 ppm at night (when consumption of CO_2 is low and respiration occurs in plants) and drops to ~300 ppm at midday at tree top level. Because the atmosphere's ratio of O_2 to CO_2 is so large (20%:0.04% = 500), fluctuations in carbon dioxide levels by factors even as high as 10 do not noticeably affect the amount of oxygen around. Nevertheless, instrumentation technology has improved to the point where minute changes in the level of atmospheric oxygen can be monitored. Shown in the figure are the results of changes observed at two sites relative to the concentration of oxygen in 1985. The long-term trend interpretation includes a straightforward recognition that increasing combustion is consuming total oxygen.

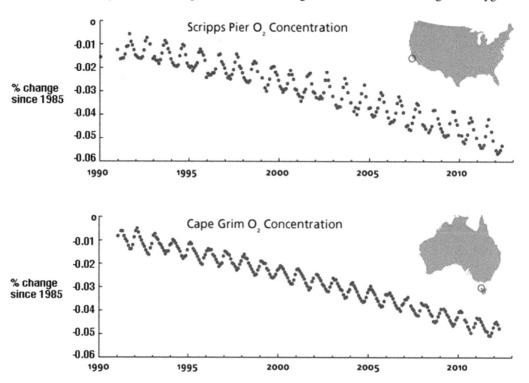

Light, temperature, and carbon dioxide levels all influence photosynthesis. Changes in the level of carbon dioxide affect the rate, of course. This is important if studying plant fossils and remnants are to be used to unravel the history of atmospheric carbon dioxide levels. Earlier, we presented the figure showing the results of a study of spinach plant growth at high CO_2 levels. (*Hydroponics*, for efficient plant growth, uses elevated carbon dioxide levels of 0.1–0.2%.) Not surprisingly, photosynthesis depends on the amount of light as well, shown by the three different curves in that figure. Complicating all this is the fact that the rate of photosynthesis varies with temperature, reaching a maximum and then declining at very high temperatures. (In contrast, respiration, which, like our breathing, generates CO_2, increases continuously with temperature.)

The combined effects of temperature and light changes show up as seasonal fluctuations quantified through their amplitudes.* Examples of this for sites in Hawaii and the South Pole were displayed in Chapter 6 ("The Air Today"). Shown here is a diagram of the difference between daytime carbon dioxide input (photosynthesis) and its night output (respiration) for a number of different sites in Siberia. There are two categories of data collection shown. The upper display, A, is for regular, "undisturbed" measurement sites, the meaning of which will become clear when we look at the lower display for "disturbed" sites. Carbon dioxide was measured weekly over a three-year period. Each of the two graphs shows the daytime intake as the top data set (above the horizontal line, with positive influx values), the nighttime output as the bottom data set below the horizontal line and the difference – or *net* uptake – as the central data set. The amplitudes of these differences vary with seasons. It also varies, but only slightly, with geographical latitude over the several Siberian experimental stations with their different temperature environments. Winter, in these northern latitudes, and even late autumn, has very low flux (flow) and so the data is understandably near zero net flux, or not even measured. Summer there has day/night fluxes nearly canceling in amplitude for the undisturbed sites measured. But in "disturbed sites" shown on the lower graph, B, day influx increases more than double. The nighttime output (negative input) by respiration remains relatively low and no longer matches the input. Net removal has about doubled. Yet growing seasons, implied by the duration of the peaks – the widths of the curves – turn out to be some 6 weeks shorter than normal. What distinguishes these sites as "disturbed?" The sites in figure B had all experienced one of the following historical perturbations during the previous decade: fire, grazing, river erosion, and vehicle activity. Very little follow-up has appeared in the literature reflecting on the influence of "disturbances" on seasonal amplitude variations of carbon dioxide levels in the atmosphere and the mechanism appears to be unresolved.[†]

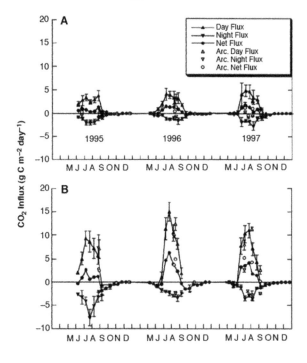

* S. A. Zimov et al., "Contribution of disturbance to increasing seasonal amplitude of atmospheric CO_2," *Science* 284, 1973 (1999).
[†] H. D. Graven et al., "Enhanced seasonal exchange of CO_2 by Northern ecosystems since 1960," *Science* 341, 1085–1089 (2013).

PHOTOSYNTHESIS ALTERNATIVES

Anaerobes are living species that operate in the absence of oxygen. They are considered to belong to two classifications. "Facultative" anaerobes can live with or without oxygen and change their metabolic chemistry accordingly. The ubiquitous *E. coli* bacteria are an example of facultative anaerobes. *E. coli* are probably the most common bacteria in the human large intestine (colon) numbering ten to a hundred billion cells per gram – and the colon typically weighs 200 g. Another type of facultative anaerobe is *clostridium* which includes in its family the *botulinum* bacteria (of "botox" notoriety) and also the bacterium that is responsible for tetanus.

But there are also "obligative" anaerobes to whom oxygen is deadly. These anaerobes are usually found in lake bottoms (and are responsible for marsh gas) and in the rumens of herbivorous animals such as cattle. They are also found in rice fields and in termites.

The presence of oxygen invariably generates some hydrogen peroxide, H_2O_2, and/or the superoxide ion, O_2^-. Normal cells produce enzymes such as *catalase* or *peroxidase* or *superoxide dismutase* to counteract these toxic substances, rendering them harmless. But obligate anaerobes don't have these protective enzymes. Unavoidable oxidation reactions produced by peroxide and superoxide are lethal to the cell. In fact, when phagocytes – those cells that help fight infection in us – engulf bacteria, much of their mechanism of disposal is to douse the bacteria with superoxide to help kill it.

"Methane bacteria" are very primitive anaerobes (either obligative or facultative) that produce methane gas from carbon dioxide according to

$$CO_2 + 4H_2 \rightarrow CH_4 + 2H_2O + energy$$

(the hydrogen being produced by other bacteria).

There are also sulfur bacteria and they, too, must live in an oxygen-free environment. The two categories of such bacteria are referred to as the purple and the green varieties and use photosynthesis. Their photosynthesis, instead of being based on $CO_2 + H_2O$, is designed around $CO_2 + H_2S$. As ordinary photosynthesis gives elemental oxygen as a by-product, the anaerobic process carried out by these sulfur bacteria gives elemental sulfur as a by-product. No oxygen is produced. Green bacteria are found near sea vents and the purple bacteria are found most often in sulfur springs. Their chlorophyll-like light receptors are different that those found in plants and algae.

Prokaryotes,* like bacteria, have populations that outnumber more complex living species. Recent estimates are that there are $4–6 \times 10^{30}$ such microbes. That's 4 to 6 million trillion trillion cells and they contain 300–550 billion tons of carbon, an amount equivalent to that in surface plants, terrestrial, and marine combined. The number of different species of prokaryotes is arguably greater than 100,000, perhaps as high as several million. Prokaryotes are found mainly in three habitats. In water, essentially the oceans, the estimate of the fraction of the total number of prokaryote cells living here is a couple of percent. In the soil, where prokaryotes are essential to microbial decomposition of organic matter generating CO_2, perhaps 5% of the population resides. Most of the prokaryote community, however, has taken up residence in the subsurface, light-free environment. This includes marine sediments where subsurface means sludge depths greater 0.1 m. For terrestrial habitats, the subsurface is considered to start at depths of 8 m. Here, populations extend down to many hundreds of meters, if not kilometers. There is indirect evidence that oil degradation by prokaryotes at depths as great as 4 km is occurring. Below this level, where temperatures are usually greater than 120°C, no life is expected to be sustainable. Oil itself is an interesting remnant of carbon dioxide processing that we will discuss in Chapter 19.

---

* Having cells whose DNA is not found within a cell nucleus.

NOBEL PRIZES RELATED TO PHOTOSYNTHESIS

The complexity of the science behind unraveling photosynthesis can also be measured by the recognition given to various scientists through the Nobel Prizes. The table here is a list of such prizes at least partially relevant to photosynthesis.

Year	Field	Name	Topic
1915	Chemistry	Wilstätter	Mg in chlorophyll
1922	Physiology/Medicine	Meyerhof and Hill	Lactic acid production
1929	Chemistry	Harden and Chelpin	Yeast metabolism
1931	Physiology/Medicine	Warburg	O_2 in tissue
1947	Physiology/Medicine	Cori and Cori	Sugar phosphates
1953	Physiology/Medicine	Lipmann and Krebs	ATP and Krebs cycle
1961	Chemistry	Calvin	Photosynthesis
1964	Physiology/Medicine	Lynen	Fatty acid metabolism
1965	Chemistry	Woodward	Chlorophyll synthesis
1970	Chemistry	Leloir	Carbohydrate synthesis
1988	Chemistry	Deisenhofer, Huber, and Michel	Photosynthesis proteins

In all things of nature there is something of the marvelous.

ARISTOTLE

15 Respiration and Metabolism

He lives most life who breathes most air.

ELIZABETH BARRETT BROWNING

The term "respiration" has a dual meaning. Most commonly, it refers to the act of breathing. But it also refers to the biological process on Earth by which oxygen is consumed and carbon dioxide is produced as a result. If respiration balanced photosynthesis, there would have been no accumulation of oxygen in the atmosphere. Burial (sequestration) of carbon dioxide derivatives prevents some respiration (decay) and enables the buildup of oxygen.

Having just finished a quick look at photosynthesis in which carbon dioxide and water combine to produce glucose,

$$6CO_2 + 6H_2O \rightarrow C_6H_{12}O_6 + 6O_2$$

we can move on to look at complementary processes, one of which is the metabolism of carbohydrates. Photosynthesis' reaction just summarized is even better expressed as a rearrangement (reversal) of this equation

$$6CO_2 + 6H_2O + energy \rightarrow C_6H_{12}O_6 + 6O_2$$

The *metabolism* of the carbohydrate can be expressed as

$$C_6H_{12}O_6 + 6O_2 \rightarrow 6CO_2 + 6H_2O + energy$$

Photosynthesis requires energy – in the form of light for our discussion. Not surprisingly, what looks like running the movie in reverse, metabolism of carbohydrate, provides energy. Metabolism is a source of energy. But as in the case of photosynthesis, the details of the metabolism of carbohydrates are complex. Yet all living things that metabolize carbohydrates using oxygen follow similar pathways.

We'll begin with a modest observation. The overall equation for the metabolism of glucose as represented by the equation looks identical to what you would write down if you were talking about burning glucose – as in a flame – in the presence of oxygen. The energy released would appear in the form of much heat, some light, and released gases (carbon dioxide and water vapor). To make use of that energy, the heat and expanding gases can be tapped. This is what an internal combustion engine and a steam engine do using hydrocarbons (not glucose). But an inescapable requirement – one that cannot be overcome by even the most ingenious engineering design – is that some of the heat is lost to the surroundings. In theory, the most efficient steam engine operating between the boiling point of water and room temperature is about 20% and this has nothing to do with insulation. The former theoretical limit is not achieved very easily. Efficiency can be improved if the hotness (temperature) of the energy source is increased relative to whatever serves at the other end as the heat-loss sink. But for living species, there aren't many examples that can tolerate temperatures above or even near the boiling point of water. What is done biologically is to use a large number of intermediate energy generating steps in such a way that the heat generated at each step is minimal, yet the efficiency of extracting the available energy is about 40%. How? That's what we'll look at.

The energy-extracting steps usually involve the combination of a molecule with the acronym ADP (*adenosine diphosphate*) with the phosphate ion (abbreviated as P⁻) to make *adenosine triphosphate*, ATP. In the very first stages of glucose metabolism, oxygen isn't needed. The pathway is referred to as the *anaerobic* sequence. A sequence of nine steps involving nine different enzymes brings about bond rearrangements, phosphate additions, breaking an intermediate six-carbon species into two 3-carbon species, and, ultimately, generating a pair of *lactic acid* molecules, $C_3H_6O_3$, whose structure is shown here.

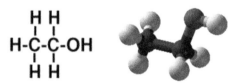

In muscle, if the anaerobic path proceeds under extreme demands, the lactic acid concentration can accumulate causing pain and the sensation of being tired. Recovery takes time. (Out-of-shape runners know this effect.) Eventually, the next part of the path brings about the concluding sequence: converting lactic acid into carbon dioxide and water. But not in all living species. In some microorganisms, yeast being a prime example, the immediate precursor of lactic acid follows an alternate path because of a different enzyme. In yeast, we refer to the metabolic process as the *fermentation* of glucose. In continuing the discussion of conventional metabolism (not that seen in yeast), conversion of each of the two molecules of lactic acid to carbon dioxide and water involves another series of steps, more than a dozen. The process is also referred to as lactic acid fermentation and has applications in the production of sauerkraut, yogurt, and kimchi. The precursor of lactic acid is another 3-carbon molecule, *pyruvic acid*, $C_3H_4O_3$.

Instead of conversion to lactic acid, each of the two pyruvic acids loses a carbon dioxide and the 2-carbon residues end up (after one more step) as *ethanol*, C_2H_6O, also known as ethyl alcohol (grain alcohol).

Production of alcohol in this fermentation is accompanied by the generation of carbon dioxide. It was this by-product from a brewery that led Priestley to his early work on fixed air. (See Chapter 3, "Discovery.")

Not yet mentioned in this simplification is where oxygen comes into play. The oxygen(s) are employed in several steps to remove hydrogen ions, regenerating some of the enzyme-helping molecules so that they are again available. In so doing, the oxygens are converted into water, H_2O. The full blown materialization, going from glucose to carbon dioxide and water is

$$C_6H_{12}O_6 + 38ADP + 38P^- + 38H^+ + 6O_2 \rightarrow 6CO_2 + 38H_2O + 38ATP$$

Where's the energy, you ask? Energy is "stored" in the ATP molecules, in certain of its bonds, from which it can be released as needed by conversion (back) to ADPs plus phosphates with lower energy in certain bonds. Storage of energy in bonds is somewhat akin to compressing a spring for later release.

PHOTORESPIRATION

In the previous chapter, where photosynthesis was being discussed, mention was made of the important enzyme *rubisco* whose main role is in speeding up "carboxylation," addition of carbon dioxide to the 5-carbon molecule abbreviated as RuBP. Rubisco is short for "ribulose bisphosphate carboxylase/oxygenase." The "-ase" suffix is what is used in biochemical nomenclature to indicate an enzyme. As with all enzymes, rubisco is a protein, a large molecule. There are more than 100 known natural variations on the structure of the molecule. RuBP is extremely inefficient with carbon dioxide, competing with oxygen for the active site (hence the split label suffix "carboxylase/oxygenase") and having sometimes only a four to one advantage over oxygen if CO_2 were as abundant as oxygen. In oxygen-poor early Earth's atmospheric history, this was no problem, but as oxygen levels grew over the eons, evolution fostered many adaptations to less friendly environments. There is more than one rubisco variety whose complete structure is known, but not all of rubisco's functions are understood. Red algae, important builders of calcium carbonate limestone reefs, have the highest selectivity for carbon dioxide *versus* oxygen, about a 200 to 1 advantage. Spinach rubisco has the structure shown here. For comparison, the relative size of carbon dioxide is approximately indicated by the barely noticeable dash in the lower right corner of the figure.

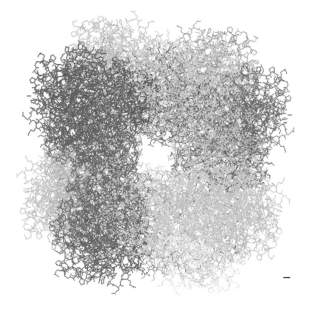

The part of its action that relates to our present discussion is rubisco's affinity for both CO_2 and RuBP and its ability to foster a rearrangement of the chemical bonds, subsequently releasing the pair of 3-carbon "PGA" molecules introduced in the previous chapter and shown again here for convenience. The rearranging of bonds is the carboxylation reaction, hence the word *carboxylase* in the enzyme's long name.

$$*CO_2 + H_2O + \underset{\text{RuBP}}{\begin{array}{c} CH_2OPO_3^{2-} \\ | \\ C=O \\ | \\ CHOH \\ | \\ CHOH \\ | \\ CH_2OPO_3^{2-} \end{array}} \xrightarrow{\text{enzyme}} \underset{\text{PGA}}{\begin{array}{c} CH_2OPO_3^{2-} \\ | \\ CHOH \\ | \\ *COOH \end{array}} + \underset{\text{PGA}}{\begin{array}{c} CH_2OPO_3^{2-} \\ | \\ CHOH \\ | \\ COOH \end{array}}$$

Rubisco has another talent in that it can also encourage the launching of a different sequence of steps. The alternative, respiration, requires light and involves the enzyme's use of oxygen, producing just one PGA molecule plus a 2-carbon by-product, *2-phosphoglycolate*, 2PG, shown here.

$$O_2 + \underset{\text{RuBP}}{\begin{array}{c} CH_2OPO_3^{2-} \\ | \\ C=O \\ | \\ CHOH \\ | \\ CHOH \\ | \\ CH_2OPO_3^{2-} \end{array}} \xrightarrow{\text{enzyme}} \underset{\text{PGA}}{\begin{array}{c} CH_2OPO_3^{2-} \\ | \\ CHOH \\ | \\ *COOH \end{array}} + \underset{\text{2PG}}{\begin{array}{c} CH_2OPO_3^{2-} \\ | \\ C \\ O^{\diagup\diagdown}O^- \end{array}} + H^+$$

The latter actually inhibits the Calvin photosynthetic pathway, leads to release of carbon dioxide, and must be metabolized for the process to ever recover. This alternative oxygenation pathway – the use of O_2 – accounts for the word *oxygenase* in the multi-function enzyme's name (or the final letter "o" in its nickname, rubisco). Requiring light to foster an oxygenation is why the process is referred to as *photorespiration*.

Because rubisco carries out its carboxylation (CO_2 addition to RuBP) comparatively slowly, lots of enzyme is required in the plant leaf. It has been found that the enzyme constitutes as much as 50% of the soluble protein in leaves, making rubisco arguably the most prevalent protein on the planet. (And its synthesis requires CO_2, as diagrammed in Chapter 14.)

On the other hand, as a result of the photorespiration pathway, one carbon dioxide is produced for each two RuBP used. Photorespiration is a drain on the plant's efficiency, taking place about a third of the time in place of photosynthesis: carbon dioxide incorporation leading to sugar. Detailed analysis of the entire photorespiration phenomenon indicates that nearly half the energy accumulated with the carboxylation and photosynthesis path is ultimately lost *via* photorespiration. The purpose, if any, of photorespiration remains mostly controversial, if not actually unknown.

Since O_2 and CO_2 compete for rubisco, the pathway tracked by the enzyme, not surprisingly, depends on the relative concentrations of O_2 and CO_2. Even if rubisco's action favors carbon dioxide use over that for oxygen by a factor of 80, the more than 500-fold excess of O_2 over CO_2 in the atmosphere now reduces that advantage to only a three-fold benefit. Photorespiration can be viewed as a feedback barrier against increasing photosynthetically derived oxygen levels. As atmospheric O_2 amounts rise due to photosynthesis, the competition with the photorespiration pathway becomes tighter. Eventually, photorespiration takes over, consuming O_2, shutting down plant growth. This puts a brake on the upper concentration of oxygen achievable.

Variations in the rubisco protein structure occur among different species. Even slight changes in structure can alter, among other things, the competition (or "selectivity") between O_2 and CO_2 for rubisco. For example, some bacteria have the CO_2 advantage reduced by factors of two or more compared with the typical factor already mentioned. This is not a problem, though, if there's no oxygen around as in anaerobic environments…or on primitive Earth. On the flip side, there are some photosynthetic systems such as those in certain heat-thriving algae that have CO_2 specificities more than twice the normal. Additionally, there are photosynthetic systems, as in many grasses, where the carbon dioxide is accumulated or concentrated in microscopic regions within a cell as an enriched feed to rubisco, restoring the advantage locally over the otherwise more abundant external O_2.

RESPIRATION AND HEMOGLOBIN

When we breathe, the inspired air contains 21% oxygen (at sea level). When we exhale, the oxygen level in the expired breath is down to 15% and there is 3.6% carbon dioxide present. The latter is almost 100 times its concentration in air. Each human exhales about a kilogram of carbon dioxide per day. That is 2.5 Gt per year for humanity.

Exhaled carbon dioxide, it turns out, can serve as an attractant to mosquitos.* They can zero in on an exhaling mammal from 100 feet away. That carbon dioxide is the unavoidable result of metabolic processes among which is the oxidation of glucose described earlier.

For fruit flies, the *drosophila*, on the other hand, it has been found that even slightly elevated levels of carbon dioxide will stimulate avoidance behavior.[†] That is the reverse of the response that mosquitoes have toward carbon dioxide. Life's tendencies evolve in interesting ways.

Oxygen entering our lungs enters into our trachea, then into the bronchial branches, and ultimately to the narrower bronchioles where everything fans out into the alveolar region strikingly displayed here.

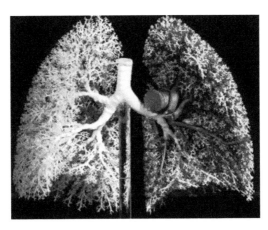

---

* M. Enserink, "What mosquitoes want: secrets of host attractions," *Science* 298, 90–92 (2002).
† G. S. B. Suh et al., "A Single population of olfactory sensory neutrons mediates an innate avoidance behavior in Drosophila," *Nature* 431, 854–858 (2004).

There are 300 million alveolar sacs in the human lung. Total alveolar surface area is nearly 100 m^2, roughly what is covered for a tennis court. O_2 diffuses across very thin membranes into capillary blood where it is taken up by a large molecular assembly called *hemoglobin* shown here.

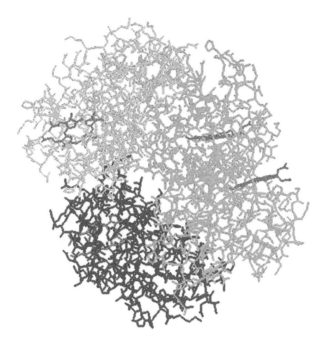

Central to the behavior of hemoglobin is a structure called *heme*. Heme is vaguely similar to chlorophyll (see the previous chapter), a conspicuous difference being the presence of iron rather than magnesium at its core. The outer regions of heme's structure do also differ from those of chlorophyll.

heme

Hemoglobin is a package of four major protein substructures, each containing one heme group. Nearly half the volume of blood is red blood cells. There are 5 million red blood cells in a milliliter of blood. Within each such cell are a few hundred million hemoglobin molecules. Blood is continuously circulated via the pumping heart and delivers the oxygenated hemoglobin to places where it is needed, such as working muscle. There, the oxygen is released by the hemoglobin. Hemoglobin enters our discussion, not because of the oxygen story line, but because of carbon dioxide as discussed next.

CO₂ AND BLOOD

Release of oxygen has an interesting and critical effect on the carbon dioxide metabolic waste product that needs to be removed. Only a minor portion of carbon dioxide diffuses out of cells as dissolved CO_2. The major portion of the carbon dioxide waste is converted to bicarbonate ion, HCO_3^-, which is very water soluble and can enter the blood and be carried back by the blood's circulation to the lungs where a new supply of oxygen is picked up at inhalation. Such a conversion process is ordinarily slow, taking many seconds or even minutes, but it is accelerated 5,000-fold by *carbonic anhydrase*, an enzyme in red blood cells. The action of taking on oxygen helps convert the bicarbonate back to carbon dioxide in the lung where it is then released as a gas during expiration. A delicate balance is necessary here. For example, where there is less metabolism and consequently less carbon dioxide waste product, the release of oxygen fuel from hemoglobin is reduced so as not to overshoot the balance. By comparison, in trying to maintain equilibrium during extra exertion, an increase in carbon dioxide is answered with increased pulmonary ventilation (breathing). This common experience had at least one unusual medical application. For a period of time beginning in the 1970s, carbon dioxide was actually used to stimulate respiration in human newborns.*

Carbonic anhydrase is an enzyme with a molecular weight of 28,000–30,000, depending on type. It was discovered in 1940 and is technically known as a metalloenzyme because it contains zinc. Carbonic anhydrase is widely distributed. In blood, three different forms are present. It speeds up the merger of CO_2 with H_2O to produce $H^+ + HCO_3^-$ (and also the reverse) by factors sometimes reaching 1-million-fold. A brief, simplified look at how the enzyme works is worthwhile.

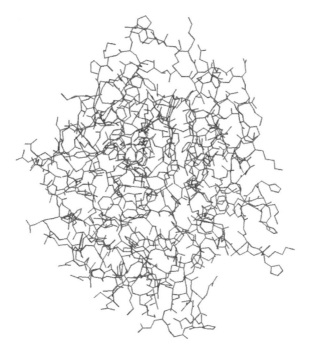

First, recall from Chapter 10, "Carbon Dioxide and Water," that a solution of carbon dioxide in water, one at moderately high pH relative to neutral, equilibrates the carbon dioxide principally as the bicarbonate form, HCO_3^-. If not in the proper pH region to do this, the zinc ion, Zn^{2+} can accomplish the same thing, behaving first as an acid by reacting with water and releasing H^+ as follows:

$$Zn^{2+} + H_2O \rightarrow Zn^{2+} - OH^- + H^+$$

---

* T. R. Myers, "Therapeutic gases for neonatal and pediatric respiratory care," *Respir. Care* 48, 399 (2003).

The enzyme carbonic anhydrase in essence uses this zinc-based reaction with water in the active region of the enzyme as a first step, "1" below. Next, "2," carbon dioxide approaches the zinc site where the OH^- attaches to the carbon on the CO_2 to be released as an HCO_3^- ion, "3" and "4." (Chapter 4 on the structure of carbon dioxide, shows that the central carbon has a partial positive charge which is naturally attracted to the oxygen negative charge on OH^-.) The role of much of the rest of the enzyme is most likely to force the carbon dioxide into the ideal geometric arrangement for this change to occur and to appropriately accommodate the original H_2O addition.

In less than a second after carbon dioxide's encounter ("2") with carbonic anhydrase, H^+ and HCO_3^- are released by the enzyme as the steps show ("His" is a *histidine* component of the protein that binds the zinc to the enzyme). The H^+ causes a drop in blood pH – an increase in acidity. But because hemoglobin can take up the H^+ and act, in effect, as a buffering chemical agent itself, the pH drops only from 7.41 to 7.37. Soluble bicarbonate HCO_3^- diffuses out of the red blood cell into the blood plasma, being replaced by ubiquitous chloride, Cl^-, because the solution must stay electrostatically neutral (uncharged). The opposite effect occurs when the bicarbonate is later released as CO_2 in the lung upon exhalation. HCO_3^- combines with H^+ released by hemoglobin when fresh oxygen is taken up by hemoglobin. The H^+ recombines with bicarbonate to form water and carbon dioxide. About 70% of the total carbon dioxide transport process follows this route. Most of the remaining carbon dioxide actually becomes loosely bonded to certain nitrogens on the hemoglobin assembly, forming a *carbaminohemoglobin* with as many as four carbon dioxides bound to one hemoglobin molecule. This is a relatively slower reaction that, through its structural influence, lowers the affinity of the hemoglobin toward oxygen. That is, in locations where there is CO_2 to be picked up (as waste in an energy consuming cell) there is a reciprocal push for more oxygen to be released. That exchange requirement is consequently met since hemoglobin's affinity for oxygen has been diminished.

The steady-state amount of bicarbonate in the blood is itself of crucial physiological importance. It serves to buffer the acidity of the blood at pH 7.4. This is determined by the ratio of HCO_3^- to "H_2CO_3" which, in blood, is about 11 to 1. (See Chapter 10, "Carbon Dioxide and Water.") Any blood whose pH is outside the range of about 7.2–7.6 can have serious medical consequences. Although there are other chemicals which also act as buffers in the blood, the predominant capacity of the blood pH to remain more or less constant is due to the bicarbonate ion. The bicarbonate buffer is not maintained automatically as an equilibrium outcome of carbon dioxide levels in the air, but rather is controlled by those amazing chemical machines, the kidneys, which can remove carbonic acid and/or pump bicarbonate back into the blood as needed.

Our rate of breathing is controlled involuntarily by the concentration of carbon dioxide in the lungs monitored by sensor cells that communicate with the *medulla oblongata* often just called the medulla. The medulla is located at the base of the brain at the top of the spinal column and controls motor nerves in the diaphragm and muscles in the rib cage. It is central to all involuntary functions such as breathing and heartbeat. Additionally, the smooth muscles in the walls of the lungs' bronchioles are sensitive to the presence of carbon dioxide. Rising concentrations of CO_2 causes the bronchioles to dilate – open wider – and this, in turn, lowers the resistance in the airways, increasing the airflow in and out.

Amphibians use their skin to "breathe" out CO_2. Frogs expunge carbon dioxide two and a half times faster through the skin than through the lungs. In contrast, humans release only 1% of carbon dioxide through the skin.

HYPERVENTILATION

Anxiety can sometimes bring about a condition known as "hyperventilation," involving rapid shallow breathing resulting in carbon dioxide being released from the blood faster than its pH can be physiologically maintained at the buffer value 7.4. Blood becomes less acidic, more basic (alkalotic), with the loss of carbon dioxide. The light-headed, dizzy feeling in "alkalosis" is often accompanied by tingling sensations in the hands and feet. Loss of consciousness can result. Hyperventilation can frequently be overcome by relaxing, breathing slowly, and, arguably, by breathing into a paper bag (so as to re-inhale higher levels of carbon dioxide and restore blood pH balance). On the other hand, self-diagnosis is not recommended since other more serious phenomena have similar symptoms. Among these are pulmonary collapse or embolism.

IT'S A CROC!

Crocodiles seize prey in their powerful jaws and then submerge. They do this to drown their victims. Crocodiles are able to remain under water for an hour or even more. This ability is enabled by a form of hemoglobin that actually reacts in a beneficial way with carbon dioxide, metabolism's by-product, in its bicarbonate form. In 1995, a team of molecular biologists* determined that the protein portion of croc hemoglobin binds with the bicarbonate ion in a place and geometry that greatly eases the release of oxygen stored on the hemoglobin, reducing the need to breathe. It acts on a molecular scale something like a surgeon's retractor does in prying structures apart.

---

* N. Hennakao Komiyama et al., *Nature* 373, 244 (1995).

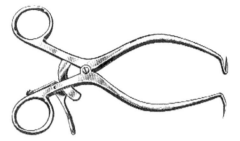

A Surgeon's Retractor

The crocodile seems to be the only species with this particular variation in hemoglobin molecular structure and its concomitant function.

APOLLO 13

Everyone is more or less familiar with the famous 1970 lunar flight of Apollo 13. Briefly, at about 47 hours following its 13:13 CST launch, Flight 13 appeared to be developing as the smoothest flight of the Apollo program. The coupled command, lunar, and service modules were on their way to a lunar landing. Nine hours later, on April 13, an electrical short circuit ignited some wire insulation. One of two oxygen tanks blew up, hopelessly damaging the remaining tank. These were the gas supplies for fuel cells that provided electricity to the launched system. Backup batteries were available for 10 hours' worth of electricity, but obviously to all, this wouldn't suffice for a longer trip. The damage from the tank explosion is clearly seen in the upper right here.

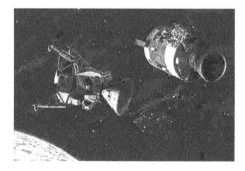

The "command module" lost its normal supply of electricity, light, and water while 200,000 miles from Earth on the way to the Moon, another 40,000 miles away. The astronauts transferred to the "lunar module" which could serve as a sort of lifeboat. There was plenty of oxygen available for breathing from a variety of sources. However, electricity use had to be cut to one-fifth normal because of damage to the fuel cells. Not enough water remained for more than five hours ordinary use, so intake of water was also cut by a factor of 5. Those were the easy challenges. The tough problem was the buildup of exhaled carbon dioxide.

Carbon dioxide was ordinarily removed from the module's air by lithium hydroxide (LiOH) canisters. The reaction is

$$CO_2(gas) + 2LiOH(solid) \rightarrow Li_2CO_3(solid) + H_2O$$

The setup in the lunar module had sufficient capacity for two men for 49 hours – the lunar landing requirement. But now it would be needed for three men for nearly 84 hours – the return to Earth.

Understandably, there were also lithium hydroxide canisters in the command module where the astronauts had been and where they were ordinarily still supposed to be, but these were of a different configuration and did not fit in the lunar module.

The carbon dioxide level in the lunar module was rising to dangerous levels. The federal standard for limit of exposure to carbon dioxide in the home is 5,000 ppm (nearly 13 times that present in the air normally). In the module, the CO_2 level was well past that and dire consequences were all but guaranteed. Air with 5% CO_2 causes rapid breathing; up to 10% causes shortness of breath, dizziness, and headaches; up to 15% results in muscle contractions and loss of coordination; higher levels lead to convulsions, unconsciousness, and ultimately death. But Mission Control in Houston, which had a replica of the modules with all the on-board supplies and equipment, devised an ingenious and elegant patch connection. Using available plastic storage bags, cardboard, space suit hoses, and adhesive tape, the astronauts were instructed on how to assemble a connection from the lunar module that would allow the air to be sucked through the other cartridges in the command module. A photo of the rigged system is seen here.

The partial pressure of carbon dioxide in the occupied module peaked at 20,000 ppm (2% CO_2) when the hookup was activated. Within an hour, the CO_2 level dropped to about 250 ppm, less than the 400 ppm we now breathe as we read this.

Near the end of the return flight, the command module was released. Shown here is an artist's recreation of the sequence of events: release of the command module, release of the lunar module which never landed on the lunar surface, and reentry of the service module into the upper atmosphere.

BREATH ANALYSIS

Valuable insight into metabolism and physiology is gained from a knowledge of carbon dioxide levels in the blood. For example, CO_2 levels indicate the balance between production and elimination. Measuring blood CO_2 is not easy. However, monitoring of carbon dioxide in exhaled breath is. The technique is not invasive and can be conducted continuously. Analysis of the CO_2 content in exhaled breath is done by determining the absorption of infrared light by the air sample. (See Chapter 4.) Monitoring carbon dioxide this way probes the severity of pulmonary problems, circulation, and integrity of breathing passages.

PLANNED UNCONSCIOUSNESS

Carbon dioxide can be used as an anesthetic and has been so used for humans, though infrequently and not in recent decades. More often it was employed with animals. It is believed that the anesthetic effect of carbon dioxide brings about a pH change in the cerebrospinal fluid and that this change induces the anesthesia. The technique is rather extreme, involving high levels of carbon dioxide, perhaps 80%, supplemented with oxygen as the remainder. The advantage is the very rapid response to the anesthetic probably owing to the ease of its chemical transport through the pulmonary and circulatory system. The disadvantage is that recovery is also quick, maybe less than a minute duration for the anesthetic effect. That disadvantage is enough to argue for its abandonment as anesthesia.

CARBON DIOXIDE FLOODING

The first open heart surgery was performed in 1953. Five years later, the use of carbon dioxide at the end of the procedure was adopted. Open heart surgery is now one of the most commonly performed surgeries. A variety of problems are solved with open heart procedures which usually involve stopping the heart for up to an hour. When the procedure is near conclusion, the heart must be closed and restarted. One of the major risks at this stage is that if any air remains within a heart's chamber, resulting air bubbles (embolisms) circulating in the blood can be fatal, especially if they reach the brain. Even microscopic emboli have resulted in documented decreased cognitive ability in postsurgical patients. Problems with air embolisms have been known as far back as the fourth century B.C. when Hippocrates commented upon it as reported in Morgagni's *De sedibus et causis morborum* (*Of the seats and causes of diseases*) in 1761. With the introduction of open heart surgery, five years of trying to address the problem of air bubbles centered on a large number of variations of mechanical procedures such as simple venting, use of needles, vacuum, and even rocking the operating table. None worked. Carbon dioxide turned out to yield a simple solution. It is introduced to flush out and displace air just before "closing" in heart surgery to prevent air embolisms. The embolism problem doesn't exist with carbon dioxide gas since CO_2 is readily absorbed by the blood unlike alternative choices such as pure nitrogen or helium.

16 Weathering

To be worn out is to be renewed.

LAO-TZE

WATER ON THE ROCKS

Weathering is the action of the elements causing disintegration. On rock, weathering is a very familiar geological process with both chemical and mechanical aspects. Almost invariably, chemical weathering involves contact with water. It is distinguished from erosion, which is mechanical wearing away of material, although erosion such as landslides should actually facilitate chemical weathering in many circumstances. Acid rain damage is one type of chemical weathering and rusting of iron is another. Washing away soil in heavy rains is an example of erosion: mechanical weathering. A superb example of erosion, illustrated in the photograph here, is the Grand Canyon in the western United States, which is as much as 5,000 feet deep, and whose formation began some 20 to 30 million years ago. Mechanical erosion is much easier to visualize than chemical weathering, but it is the latter where our current interests lie although the two versions of weathering are interconnected.

CO$_2$ SPONGE

Weathering is essential to understanding the journeys of carbon dioxide in nature over the eons. It involves the long-term interplay between sources and sinks that stabilizes or disrupts the legacy of carbon dioxide. Chemical weathering nowadays is estimated to account for greater than 50% of the sequestered carbon dioxide per year.*

Chemical weathering understandably must depend on the availability of water. Water not only penetrates rocks, but also carries "corrosive" chemicals into intimate contact with them. Percolating rainwater results in continued dissolving of soluble weathered mineral products in fresh water and transporting them elsewhere, to streams, rivers, and the ocean, or below ground into caves. Arid regions clearly weather slowly. Contact with water depends not only on rainfall, but on topography – does the terrain slope gently or steeply? – and other conditions as well. Of course, if minerals in rocks are covered by ice or glaciers or even soil then weathering won't occur and carbon dioxide will not be consumed from the atmosphere.

Mountains are major actors in this phenomenon. For example, it would seem that the vast quantities of dissolved rock carried by the Amazon River and its many tributaries originate in the expansive low-lying Amazon basin where contact with river water is prolonged. But most of the dissolved substances originate in the Andes Mountains which are easily weathered and which, over time, present fresh minerals to be weathered as the mountain chain (on the "ring of fire," see Chapter 13) continues to rise. Indeed, it is estimated that roughly half the weathering (chemical and mechanical) occurs on the steepest 10% of the planet's topography.[†]

Weathering can be slow, depending on temperature, composition, humidity, grain size, and rock fracturing. Some minerals swell upon their chemical "hydration" with water. Depending on the mineral, the expansion can amount to as much as three times the volume of the un-hydrated material. Mechanical penetration by plant roots into rock assists in the weathering process. Roots anchor soil in place, slowing down possible erosion thereby keeping soil and water in contact for longer periods. Modern day trees of some species have root lengths that are in excess of 60 m. Eventually, roots decay, forming channels allowing deeper water infiltration. Decay also forms decomposition products, including organic acids that penetrate deep into rock where mineral dissolution will occur. Decaying organic matter from trees and other land plants fosters bacterial growth that contributes carbonic acid to further weather minerals. The soil around vegetation roots is generally found to be more acidic than surrounding soil. Naturally, these processes require the presence of terrestrial plant life, unknown a billion years ago. Biology is now an important input factor to geology.

Mechanical weathering from freeze–thaw cycles of water itself also contributes to shattering rock structures in cooler terrestrial regions. On a global basis, weathering amounts to an average loss of about 0.1 mm of solid terrain per year or 1 cm per century. For geological stretches, this is identical to a thousand feet every few million years.

DISCOVERY

The role of weathering of rocks and its underlying causes seems to have been first appreciated by the French chemist, Jacques-Joseph Ebelmen, in 1845.[‡] Ebelmen postulated that silicate – an example being the feldspar *albite* ($NaAlSi_3O_8$), an abundant chemical in crustal rocks – would consume huge quantities of carbon dioxide, releasing water-soluble bicarbonate (HCO_3^-), and leave behind silica, SiO_2, and a clay-like mineral such as *kaolinite,* $Al_2Si_2O_5(OH)_4$.

---

* O. Jagoutz, F. A. Macdonald, and L. Royden, "Low-altitude arc-continent collision as a driver for global cooling," *Proc. Natl. Acad. Sci.* 113, 4935–4940 (2016).

† I. J. Larsen, D. R. Montgomery, and H. M. Greenberg, "The contribution of mountains to global denudation," *Geology* 42, 527 (2014).

‡ J. J. Ébelmen, "Sur les produits de la decomposition des espèces minerals de la famile des silicates," *Ann. Res. Mines* 12, 627–654 (1845).

Jacques-Joseph Ebelman can legitimately be recognized as the founder of weathering geoscience. In 1845, he published fundamental factors that he regarded as affecting the levels of carbon dioxide *and oxygen* over many millions of years. His insight to the various factors was not only essentially complete, compared to what we now know, but also error-free. A careful experimentalist, Ebelman studied the chemical reactions and analyses of rocks and their *in-situ* weathering products. Samples were taken from multiple locations, a step recognized today as very necessary for scientific method validation. Ebelman measured how weathering products were lost to solutions, in particular limestone to (carbonic) acid and to sulfuric acid formed from iron pyrite weathering. He did all this well before Tyndall and also Högbom (Chapter 6, "The Air Today") hypothesized their ideas. To quote Robert Berner's translations of Ebelman's prescient conclusions,* we have the following thorough, cogent précis:

The action of organic matter, either during the growth of vegetation or accompanying its decomposition contributes probably to the weathering of silicates. One knows that certain elements from minerals, silica, alkalis, alkaline earths, iron and manganese are essential for the constitution of vegetation.

One can admit that the roots (of vegetation) can produce or accelerate the weathering of silicates with which they are in contact. On the other hand, the decomposition of organic materials in the soil exercises, as we have already seen, a dissolving action on many of the materials, which enter into its decomposition, particularly on the ferruginous constituents, and it, is probable that some organic acids other than carbonic acid contribute to this reaction.

The terrestrially derived (dissolved) carbonates end up by being deposited or they are taken up by marine animals, molluscs, and zoophytes. The alkali carbonates react with the calcareous salts contained in seawater, precipitating a proportional quantity of calcium carbonate.

I see in volcanic phenomena the principal cause that restores carbon dioxide to the atmosphere that is removed by the decomposition of rocks.

---

* R. A. Berner, "Jacques-Joseph Ébelmen, the founder of earth system science," *C. R. Geosci.* 344, 544–548 (2012).

Jacques Ebelman became aware of the need for a near-equilibrium relationship between carbon dioxide draw down by silicate weathering and its release by volcanic activity. He recognized that the equilibrium was not going to be perfect and that fluctuations could be reflected in changes in carbon dioxide levels (and oxygen levels as well) over time. Ebelman speculated that such changes could affect climate. He hazarded that the cessation of volcanic activity, a major contributor to the carbon dioxide balance, would promote the near complete removal of carbon dioxide from the atmosphere leading to the disappearance of vegetable and animal life. This remarkable insight occurred one and a half centuries ago.

WEATHERING AND CARBON DIOXIDE

Bicarbonate in river waters comes mostly from chemical weathering of minerals. Examples involving solid calcite and also a feldspar silicate, similar to what Ebelmen studied in the mid-1800s, are given here. Recognize that these particular reactions also mobilize calcium ions or sodium ions.

$$CaCO_3 + H_2O + CO_2 \rightarrow Ca^{2+} + 2HCO_3^- \ (\text{carbonate weathering})$$
$$2NaAlSi_3O_8 \ (albite) + 2CO_2 + 3H_2O \rightarrow 2Na^+ + 2HCO_3^- + Al_2Si_2O_5(OH)_4 \ (kaolinite) + 4SiO_2$$

Carbonate weathering, the first reaction here, consumes carbon dioxide and produces both Ca^{2+} and HCO_3^- ions. In essence, it moves calcium carbonate from one place (exposed limestone and other minerals) to another (ocean beds) as calcite or aragonite precipitate or from shells as discussed in Chapter 12. Aragonite is estimated currently to have an annual global deposition rate from carbonate weathering of 0.8 Gt. The specific silicate weathering reaction shown in the second reaction, in contrast to carbonate weathering, draws down carbon dioxide for bicarbonate marine dispersal.

Most of Earth's crust is made of rocks and much of the rocks' actual composition is a silicate of one form or another. Besides the aluminum-containing *albite*, another common feldspar contains calcium. This allows us to greatly simplify silicate weathering, bearing in mind the familial relationship between carbon and silicon indicated in the Periodic Table (see Chapter 1). Silicate weathering represented simply by the conversion of calcium silicates to limestone, is a process whose influence was emphasized by the American Nobel laureate Harold Urey.

$$CaSiO_3 + CO_2 \rightarrow CaCO_3 + SiO_2$$

This consumption of carbon dioxide, and its reverse, the metamorphosis of carbonate sediments by *heat*,

$$CaCO_3 + SiO_2 \rightarrow CaSiO_3 + CO_2$$

are collectively known as "Urey reactions." Both depend on temperature. Under cool situations, more carbon dioxide dissolves in water. An obvious outcome of this is that if atmospheric carbon dioxide levels are low and temperatures have cooled, even more carbon dioxide will be taken up into aqueous (marine) systems. Water in equilibrium with atmospheric carbon dioxide and having no other source of CO_2 can dissolve up to 0.6 milligrams of that gas per kilogram: ~5 pounds in 1 million gallons as discussed in Chapter 10. The ocean covers 70% of the surface to an average depth of 4 km. If that were pure water in contact with nothing other than atmosphere, the total carbon dioxide dissolved could amount to 1.2 megatons assuming thorough mixing. The atmosphere has 3 million megatons of CO_2.

By now it should be apparent that the bookkeeping challenges associated with carbon dioxide are quite complex. But this is even more so when it is recognized, as has been the case since the late 1970s, that there is yet another major accounting category to consider. This is the reaction of silicates on and beneath the seafloor as warm sea water permeates the rock and exchanges its dissolved

CO_2 following the Urey reactions resulting in the deposit of solid carbonates, mostly calcite, into the rocks. Estimates are that the drawdown of CO_2 by this process is comparable to the outgassing of CO_2 at ocean vents but much more study needs to be done to quantify the phenomenon further.

EARLY EARTH

During the earliest period of our planet's existence, called the Hadean Eon, continual bombardment by comets, meteors, and even large asteroids ejected impact material that would be pulverized, glassified, or structurally damaged in other ways that would subject them to easy chemical attack: weathering. Just how intense this was can be reasonably estimated by looking at the scarred surface of both the moon and the planet Mercury where the typical impact mass for a large basin can be approximated as 100 trillion tons. Some impacts were even more massive. Since Earth presents a target that is about 20 times larger than the moon, the number of impact craters has to be scaled up to get an idea of how Earth was peppered. A growing consensus pictures the probable high level* of carbon dioxide in the earliest atmosphere as being rapidly extracted by silicate weathering facilitated by abundant extraterrestrial bombardment. (The latter, by its nature, is <u>not</u> mechanical weathering.) By "rapid" is understood to mean tens of thousands of years…or more.

LIFE AND WEATHERING

Microbes closely associated with roots, and the roots themselves, undergo respiration releasing carbon dioxide into the moist rocks and soil. This combination attacks local silicates, producing Ca^{2+} ions and bicarbonate ions which are most often carried by water runoff eventually into the oceans.

Vascular plants are those that extensively incorporate *lignin*. They include trees, of course, but also ferns, flowering plants, and what we commonly call shrubs. Terrestrial vascular plants, evolving 450 Mya, intensely accelerated weathering of silicates by the vegetation's deep rooting capability. Evidence for presence of deep roots as far as 390 Mya was determined from the ages of the strata they are found in. The factor by which weathering increased starting about that time may have been between ten and a hundred. Weathering worked its way slowly over millions of years resulting in substantially lowered atmospheric CO_2 levels 400–360 Mya. In one study in North Carolina, a forest plot was intentionally exposed to increased levels of carbon dioxide, to as high as 570 ppm compared to the atmospheric level then of 270 ppm. Of course, that larger amount of CO_2 doesn't approach what is thought to have been present during the Devonian Period (see Chapter 7). The experiment revealed that dissolved bicarbonate ions increased 37% during the elevated CO_2 study. But this was for a limited collection of woody plants and conducted over only a few years of time. An illustration[†] of the atmospheric levels of carbon dioxide through the Devonian Period relative to the present day amount is shown on the next page. The "mean" curve with square data points traces the trend in results from microalgae, leaf pore indexes, and soil ^{13}C and ^{11}B proxy measurements[‡] binned into 20-million-year increments. Note the distinct drop from the right interpreted as due to the rise in plant consumption of CO_2 around 300 Mya. The production of lignin by woody plants contributed to the drop. Lignin is not rapidly degraded by microbe action and its burial in sediments on land led to large quantities of coal beds and the net diminution of atmospheric carbon dioxide.

---

* Perhaps more than 10 atmospheres *versus* 0.0004 atmospheres nowadays.
[†] R. E. Zeebe, "History of seawater carbonate chemistry, atmospheric CO_2, and ocean acidification," *Ann. Rev. Earth Planet. Sci.* 40, 141–165 (2012).
[‡] D. L. Royer et al., "CO_2 as a primary driver of Phanerozoic climate," *GSA Today* 14, 4–10 (2004).

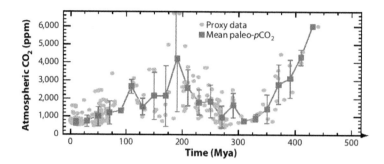

Intricate relationships develop among geologic, atmospheric, and biologic processes. As CO_2 is consumed by weathering – weathering enhanced by tectonic activity such as the rise of the Himalayas and Andes over the past 50 million years – large woody plants probably responded in a somewhat surprising way. Abundant in forests, large woody plants could end up being significantly undernourished, if not starved, as carbon dioxide levels were drawn down toward 200 ppm and perhaps lower. Trees would lose their competitive edge to plants such as grasses, plants that are much less aggressive in weathering rocks. This change in vegetative domination would serve to put a brake on the plunge in carbon dioxide's abundance in the atmosphere, a reasonable conjecture.

The next figure shows carbon dioxide levels in the atmosphere over the Eocene, Oligocene, and Miocene Epochs covering the past 50 million years. The trend is inferred from proxies such as the plant biochemical markers alkenones (shaded gray, see Chapter 8) and isotope fractionations in boron (open circles, see Chapter 8). Major tectonic land uplifting episodes for the Himalayas and Andes mountains are indicated within the horizontal bands. Note that carbon dioxide levels were presumably as high as 2,000–2,500 ppm over some part of the Eocene Epoch. Levels plunged to ≈200 ppm in the Miocene during which the evolution of modern terrestrial plants and animals seems to have occurred.

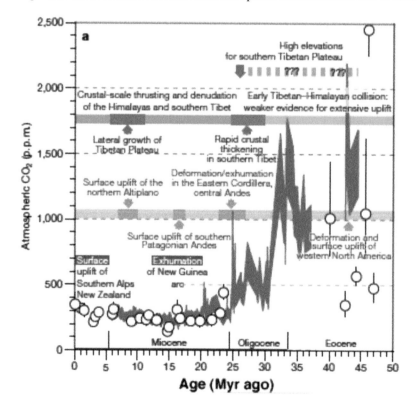

It appears from this data that the carbon dioxide levels were relatively constant over much of the Miocene Epoch. But more detailed studies of plant stomata, the number of pores on a given area of leaf, indicate meaningful fluctuations about what looks like a flat bottom. The number of leaf pores is found to increase when atmospheric CO_2 levels decrease, in that way compensating for the scarcity of sustaining CO_2 and *vice versa*. The trend is not only logical, which makes it tempting to tap into, but has also been verified by experiments under controlled conditions on several modern plant species reasonably thought to have existed through the early Miocene as well.

A 2008 study of dated fossil leaf samples from Central Europe showed the carbon dioxide levels plotted in the graph here. Ages to which the fossil leaves were assigned were determined by vertebrate fossil and magnetic orientation measurements from the same strata containing the leaves. The "mid-Miocene climate optimum" at nearly 500 ppm carbon dioxide and the other fluctuations indicated by the fossil studies summarized in the figure suggest changes that likely stressed both animal and plant survivability and suitability during this period. The evolution of modern mammals like horses, camels, antelopes, and many familiar others has been associated with such environmental stresses.

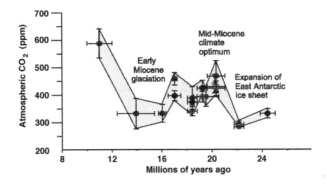

YET ANOTHER CONSIDERATION

In 1845, Jacques Ebelmen, the founder of weathering geoscience, had already considered the probable influence of iron minerals on the carbon dioxide balance. Yes, iron, Fe, element number 26. A relatively abundant iron mineral, *pyrite*, also known as fool's gold because of its color and shine is shown here. Pyrite, FeS_2, is one-quarter as dense as gold.

Pyrite is the crust's most abundant sulfide mineral. If exposed to moisture in the presence of air (oxygen), sulfuric acid can be slowly produced from the sulfide. The sulfuric acid is very aggressive in reacting with limestone to release carbon dioxide directly as in the following pair of reaction steps

$$4\,FeS_2\,(pyrite) + 15\,O_2 + 14\,H_2O \rightarrow 4\,Fe(OH)_3 + 8\,H_2SO_4$$

$$CaCO_3 + H_2SO_4 \rightarrow CO_2\,(g) + H_2O + Ca^{2+} + SO_4^{2-}$$

The significance of pyrite's role is confounding. Tectonic promoting of newly exposed, weatherable mountain terrain can contain not only silicates to draw down carbon dioxide but also can contain sulfides to resupply CO_2.

MORE RECENTLY

Despite all the complexities associated with weathering and its contribution to the atmospheric carbon dioxide seesaw pattern, major CO_2 stimulants such as weathering and volcanoes have kept levels in apparent moderate balance over the past $\approx$800,000 years as evidenced by ice core samples analyzed for their trapped air CO_2 content.* Carbon dioxide apparently varied relatively slightly, compared with the more distant past, at 200–300 ppm. (The horizontal dashed lines represent mean values over successive large time intervals.) Of course, resolving variations over a few millennia as evident in the figure here cannot (yet) be accommodated with more ancient data and so the existence or absence of near constancy cannot be established. It is undetermined. There is a very strong correlation, though, between the peak carbon dioxide concentrations and evidence that their times matched with interglacial (relatively warm) periods.

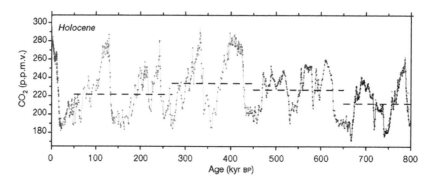

UH-OH, SINKHOLES

The re-dissolution of carbonate minerals plays not only an important role in the global scheme of carbon dioxide cycling, but also in local phenomena. When subterranean limestone dissolves in acidic water, the disappearance of underground support structure can have incredibly theatrical effects as evidenced by the photograph of a sinkhole in Guatemala City, shown next. The phenomenon is a known risk in limestone-rich areas of the world. That sinkhole volume pictured is about 3,000 m³ which, if limestone, corresponds to a content of a bit over 3,000 tons of carbon dioxide flushed away. Sinkholes are also known as cenotes, swallets, dolines, and tiankengs. Mammoth Cave in Kentucky is the longest cave in the world with some 360 miles of passageways. It has a frequently visited sinkhole called the Cedar Sink in one corner.

---

* D. Lüthi et al., "High-resolution carbon dioxide concentration record 650,000–800,000 years before present," *Nature* 453, 379–382 (2008).

The deepest water-filled sinkhole, to date, is in Tamaulipas, Mexico. The *cenote* (cenote is Spanish for water-filled sinkhole) has a depth of over 300 m, roughly the height of the Eiffel Tower. That sinkhole, the Zacatón Cenote, has been dated back to the Pleistocene. The weathering mode of formation of this sinkhole follows a process that is the inverse of the usual route. Volcanic activity below the surface perfuses the emanating carbon dioxide into the sink's water, acidifying the water, which then dissolves the limestone environment. It's much like a giant dental cavity.

For water continually dropping will wear hard rocks hollow.

PLUTARCH

17 Carbon Dioxide and Ice Overs

Some say the world will end in fire,
Some say in ice

ROBERT FROST

The past quarter century has seen some significant changes in our understanding of the behavior of the planet's surface over the course of its history. New developments involve an assortment of investigative efforts by a variety of scientists, experimental and theoretical, interested in geology, geochemistry, and biogeochemistry. Putting the bottom line first, there is growing confidence in a narrative where hundreds of millions of years ago, extended ice ages with average temperatures as low as −40°C (−40°F) and lasting millions of years encrusted most, if not all, of the planet, nearly shutting down life. These frigid spans were followed by periods, again persisting for millions of years, where the planet became sauna-like with temperatures perhaps as high as +50°C (122°F). We are talking average temperatures! Understanding such cycles forces us to bundle together most of what has been introduced in the previous chapters. There is a single explanation and it is fascinating in the way it addresses an incredible multitude of observations and puzzles intimately tied to carbon dioxide. The premise is called "snowball Earth," a term coined in 1992.

SNOWBALL EARTH

Dramatic as it certainly seems, the picture behind snowball Earth is that our planet went through a number of episodes in which it was basically entirely covered in ice to depths of hundreds of feet. How could this happen? What were the consequences? More to the point, how do we know? And what has this to do with the legacy of carbon dioxide? An abbreviated discussion follows. But a much more thorough review can be found in the 2011 "Climate of the Neoproterozoic."*

Let us talk about glaciers first and the evidence they provide us. Glaciers flow, sometimes slowly, 1 m per day (hence the adjective *glacial*), sometimes quickly, 25 m or more per day. Sometimes they melt and retreat. Yet melting can actually cause a glacier to lunge forward in certain circumstances, somewhat akin to what happens when you remove a barricade that had been restraining a pressing crowd of people. Advancing glaciers gouge out valleys, crushing and carrying quantities of rocks along the way: an extreme mode of mechanical weathering. A glacial valley in the Himalayas is pictured here.

---

* R. T. Pierrehumbert, D. S. Abbot, A. Voigt, and D. Koll, "Climate of the Neoproterozoic," *Ann. Rev. Earth Planet. Sci.* 39, 417–460 (2011).

If they reach the sea, glaciers can dump their accumulated sweepings into the waters where the detritus settles to the bottom. If glaciers "calve," yielding icebergs, these can float away from the shoreline even hundreds of miles where they likewise can disgorge debris.

Glacial debris can be recognized in sediments that had once been submerged under the ocean. "Ice rock," stones with their scrape marks – parallel grooves imprinted by the unstoppable ice flow – are among the more obvious clues. Around such stones at the bottom of the sea, fine silts fill in gaps between larger pebbles. Some such fragments as large as boulders are appropriately called drop-stones. Eventually the silt serves to cement the deposits into a solid layer over time. If, eons later, they happen to be lifted again by crustal motion, they might be discovered by curious eyes. Layers of re-lifted ice rock can be as much as 2 miles thick indicating very long periods of time for their accumulation. In the picture , a dropstone can be distinguished by the indentation in the sediment made when the stone descended into soft sediment, dimpling the existing mushy silt whose record chronicles the structure today. The distorted layer's shape under the stone minimizes the likelihood that the stone rolled or slid into place from a distance.

There are numerous places where ice rock layers have been found across the globe, all about the same approximate age, dating toward the Neoproterozoic Era.* This is the younger (more recent) boundary of the Precambrian more than 550 million years ago. Now the evidence, the pieces of the jigsaw, gets more interesting. The glacial remains in these cases present the puzzling situation of being sandwiched between two limestone layers. Wherever they are found, on any continent, the bottom limestone calcium carbonate layers are enriched in ^{13}C. The capping layers are depleted in ^{13}C. Depletion of this heavier isotope of carbon is a strong indication of absence of biological (photosynthetic) action arguably more recent than when the ice-scarred rock layers they bury were formed.

The enigma here has several aspects to it. First, the ubiquitous layer of glacial deposits – "ice rocks" – are found far from polar regions where glaciers are now dominant. What are ice rocks doing near equatorial regions? Second, the carbonate limestones are physically of the type indicating their precipitation from warm ocean water, directly contradicting the idea that the sandwiched ice rock sediments had been accumulating in glacial environments. Some of the limestone deposits are of aragonite crystals nearly 2 m in thickness. Aragonite, as the earlier Chapter 12 ("Carbonates") indicated, is the form of calcium carbonate that will be the first to precipitate from saturated calcium carbonate solution because of its lower solubility.

Further complicating the picture is another feature of these deposits: red ironstone. Red ironstone, visible as bands in the next picture, is a marker or proxy for the switch from low oxygen levels to high oxygen levels in the ocean. How does this come about and why is it relevant?

_____

* 1,000–541 Mya (see Chapter 7).

banded iron formation sample from the Soudan Iron Formation, Minnesota. x2.
(Collected by PK Strother, 1974)

Iron ions (Fe^{2+}, also symbolized as Fe^{II}) from undersea volcanoes, mid-oceanic ridge formations, and from the weathering of terrestrial rock by acidic waters, are removed from solution in the presence of oxygen by conversion to the oxide's insoluble red Fe^{3+} (Fe^{III}) form. When oxygen levels are low, however, the Fe^{2+} accumulates in such "anoxic" solutions and doesn't yield red deposits. Any appreciable subsequent appearance of oxygen in the atmosphere or surface waters such as *via* flourishing photosynthetic bacteria consuming carbon dioxide, would then efficiently precipitate the iron, giving rise to sedimenting red iron oxide bands. Yet at the time of these iron band formations, oxygen was already abundant in the atmosphere (see Chapter 7). That means that finding the red bands is inconsistent with the fact that atmospheric oxygen was already sufficiently plentiful at the time of sedimentation in the Precambrian. Oxygen should have prevented the growing concentrations of the dissolved Fe^{II} to levels that ultimately would be responsible for accumulations of banded iron deposits...unless...unless... the atmospheric oxygen could not get to the ocean environment. Was there a barrier? Ice?

In sum, to get the red bands, large accumulations of Fe^{II} must occur in the absence of marine oxygen, followed some time later by a large increase of *available* oxygen for conversion to solid Fe_2O_3. The scene is further complicated by layering where the deep ocean can be anoxic but near-surface water layers can be oxygenated.

Here is the interpretation of this collection of observations, the assembling of the puzzle pieces. The explanation itself has evolved over the past several years but started in the 1960s. W. B. Harland and M. J. S. Rudwick published an article in *Scientific American* titled "The Great Infra-Cambrian Glaciation" proposing a major ice covering. Almost simultaneously, M. Budyko from Leningrad's Geophysical Observatory also published results of calculations predicting that, because of albedo (reflection) effects, if ice coverings ever reached a point-of-no-return below 30°N/S into the tropics and equatorial regions, the entire planet would subsequently freeze over permanently. (The entire continent of Africa extends from 37°N to 34°S.) All life would be extinguished. In fact, that problematic conclusion initially led to the hypothesis that total coverage glaciation did not occur.

The current perspective on this proposed catastrophe is more complex. Look at the conventional scheme. Glacier growth, nourished by snowfall accumulations, is understandably constantly spreading ice flows outward from the colder polar regions toward the temperate and tropical regions, all other things being equal. What stops them ordinarily (like nowadays) from extending without limit besides their melting? As glaciers spread over more and more land surface – in Canada, Alaska, northern Europe, and Siberia – rocks that previously had been available for chemical weathering become buried under ice. The rocks, remember, with their abundant silicates and carbonates, are major removal venues – sinks – for carbon dioxide. As seen earlier (Chapter 16), for calcium carbonate

$$CaCO_3 + H_2O + CO_2 \rightarrow Ca^{2+} + 2HCO_3^-$$

and for calcium silicate

$$CaSiO_3 + 3H_2O + 2CO_2 \rightarrow Ca^{2+} + 2HCO_3^- + H_4SiO_4$$

That is, carbonate and silicate minerals react with atmospheric or dissolved carbon dioxide to produce ions that can be carried away in solution as runoff, for example. Minerals containing calcium carbonate and/or calcium silicate will slowly dissolve in water containing CO_2 producing soluble bicarbonate and also calcium ions that are transported into the oceans and seas. The result of glaciers blocking more landmasses is that carbon dioxide gradually accumulates in the atmosphere since a major removal route has been subdued. CO_2 is continually being fed to the atmosphere by volcanic activity. With accrual of carbon dioxide, the greenhouse effect slowly raises global temperature. Glaciers retreat, re-exposing surface material that will reset the weathering chemistry for another cycle. This description* is essentially what pertains to conventional ice age cycles with their associated centuries of interglacial warm periods such as the most recent 12,000 years.

However, geologically slow tectonics change the longer-term picture. Those weathering landmasses mentioned above that are assigned a critical role in recent eras of glacial comings and goings have been drifting over the eons as the crustal plates move around. A remarkable contemporary view of the plates and ridges is evident.

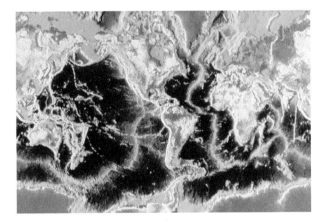

Geographies of landmasses being influenced by mid-oceanic ridge forces, for example, shift constantly but slowly. The idea with snowball Earth is that some 590–750 Mya, basically *all* the landmasses found themselves clustered near equatorial latitudes with none near the polar regions. By 900 Mya, a tropical supercontinent, named *Rodinia* by geologists, existed. (See the global map.)

* Ignoring the so-called Milankovic orbital oscillations of Earth's solar system path which affects solar input.

Its subsequent break-up over 700 Mya was preceded by flood volcanism, if not perhaps even triggered by that phenomenon. Evidence for the flood volcanism is basaltic flood deposits found in south China and dated to this very era. Left behind after the eruptions, the massive volcanic flood deposits extremely rich in silicate minerals as lava is, would have been particularly efficient in removing atmospheric carbon dioxide by weathering chemistry in the tropics.

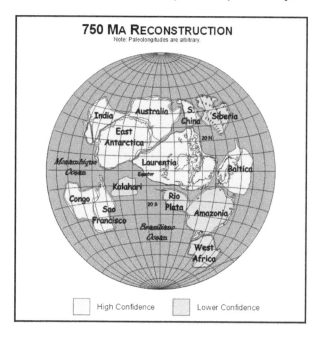

Climate modelers have proposed atmospheric carbon dioxide levels to have been at 1,830 ppm at that time, that is, nearly five times today's figure. Ordinarily, the reader's response to that situation would be "global warming greenhouse effect." However, also back that far, the amount of solar radiation was less than nowadays, a difference of 5–6% is frequently quoted. With reduced solar input, the high CO_2 levels were insufficient by half to warm things up to today's 15°C average.* How does this change the plot?

RUNAWAY GLACIATION

Climate calculations suggest that the mean global temperature at that time, 590–750 Mya, may have been cold: 11°C. As glaciers descended from polar regions, reduction in carbon dioxide removal by rocks weathering was unaffected since northern latitude landmasses didn't exist. Landmasses were all near the equator. Polar ice spread further and further. Ice has a far greater efficiency in reflecting sunlight than does land, so as the ice spread, more of the incoming solar energy was radiated away, a process that simply encouraged further cooling and growth of the glaciers. Temperatures continued to drop. Eventually, according to climate calculations, the carbon dioxide level dropped to nearly one-quarter of its earlier values and temperatures approached freezing. Oceans got ice covered. Geometrically, most sunlight "registers" near the equator, the region of the planet that points more or less directly at the sun (as vacationers and residents in the tropics know). The closer the glaciers crept to the equator, the greater was the proportion of sunlight reflected away. We have a feedback trend here abetting the spread of the ice veneer and allowing temperatures to drop more and more.

---

* To have kept temperature at today's global average, carbon dioxide levels in the atmosphere would have to have been greater than 3,000 ppm.

Conceivably, the entire planet could have been engulfed in a crust of ice pack as much as a mile thick providing unfavorable Antarctic-like temperatures as low as −50°C and the Arctic at −100°C. Of course, as the last remnants of the ocean get covered, the source of water vapor that would condense to snow, nurturing glacier evolution would be nearly eliminated although some sublimation of water vapor from ice would continue. An average temperature of −50°C does not preclude local polar temperatures below −78°C at which point carbon dioxide freezes out of the atmosphere (snows/sleets?) as observed recently for the planet Mars's polar ice caps. This would be yet another possible sink for carbon dioxide during the snowball Earth cataclysm.

At first glance, the development just hypothesized and modeled could be viewed as beyond credibility. As an end game, it is inconsistent with all of us being here to argue the point. Another piece of the puzzle: it implies all engines stop! The point of no return. Weathering induced by rainfall runoff would be minimal. Sedimentation processes even cease, minimizing geological deposits and their proxy recordings. If the oceans are covered, dropstones are no longer produced. Unlike the present day oceans, stratification – layers – would be absent and there would be exchanges between deep waters and surface waters. The inferred finality of snowball Earth evoked a lot of skepticism since clearly this ultimate encapsulation could not explain the present proliferate life. Nevertheless, the concept fed a Hollywood (unscientific) spectacular film in 2004 in which a similar, many-millennium change was accelerated for theatrical effect to occur in a matter of days.

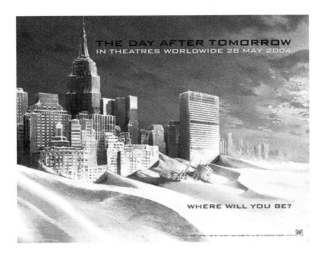

RUNAWAY SAUNA

Continuing with the snowball Earth saga, we have portrayed everything as left entombed in ice. But not forever, only for millions of years. There is no supply of oxygen to the oceans from the atmosphere. The atmosphere and oceans are decoupled. There is no physical or chemical communication between them on first thinking about the arrangement. Ice covering the equatorial landmasses would probably end up relatively thin since snowfall accumulations could no longer be called upon to replenish the glacier masses that slowly flowed outward. Ocean currents and winds diminish. Eventually, microbial activity leads to consumption of the last of dissolved oxygen.

Submerged volcanic activity does not generate enough heat to melt the glaciers other than perhaps locally. But volcanoes do release chemicals into the marine environment, many of them metallic ions. Among the ions would be iron and manganese (element number 25), soluble in oxygen-depleted seawater as the "ferrous ion," Fe^{2+}, and "manganous ion," Mn^{2+}, that can slowly accumulate in the anoxic sea water environment. Oceans become enriched in all the minerals that would cause bacterial life to flourish if the remaining needs were provided as well. Also liberated by volcanoes is carbon dioxide, produced, as we have noted before, from the metamorphic decomposition of sub-ducted carbonates at high temperature and pressure (unaffected by frigid surface conditions):

$$Heat + CaCO_3 + SiO_2 \rightarrow CaSiO_3 + CO_2$$

Carbon dioxide gas can slowly seep back into the atmosphere through any remaining openings in the ice, through cracks and fissures. The prevalence of such escape routes is important in consider-ing how thorough the extent of glaciation was. To give you a handle on the significance of this, the total amount of carbon dioxide being released to the atmosphere by metamorphic processes is cur-rently proceeding at the rate of 0.03 atmospheres worth of carbon dioxide per million years, some to the atmosphere and some into the ocean. That is, if no removal mechanisms (sinks) existed and there were no other sources of carbon dioxide, the current level of 0.0004 atmospheres of carbon dioxide (400 ppm) in the air would increase more than 70-fold over 1 million years. Since photosyn-thesis would have ceased, the carbon dioxide would return to a $^{13}C/^{12}C$ level of −0.6% representing primal volcanic composition relative to the standard enrichment ratio.

The quoted model estimate for snowball Earth is that after at least 4 million years of green-house gas accumulation, the CO_2 level would climb to 350 times today's concentration. Some[*] have argued that this level of carbon dioxide would not raise the global temperature enough. But then even longer-term outgassing would provide yet more carbon dioxide and merely extend the snowball ice age to as much as 35-million-years duration. Furthermore, ancient volcanic eruption frequencies are not yet well understood and could reasonably have been much higher than evidenced recently (or lower, but that seems to be arguable). Atmospheric CO_2 concentration could have risen to as much as six-tenths of the total atmosphere's content. The greenhouse effect would raise the temperatures ini-tially at the equator, unquestionably causing glaciers to recede. Oceans and the atmosphere would no longer be incommunicado. The reflection of sunlight would be less effective with the ice vanish-ing thus allowing further warming. During the warming, water would evaporate and water vapor, as another greenhouse gas, would accelerate the change. Deglaciation and warming calculated to eventually ricochet to +50°C would happen much more rapidly than the reverse glaciation process. Violent storms would wash the planet with carbonic acid laden rainwater, initially at the warmest "tropical" equatorial regions where the landmasses were located, refueling chemical weathering, and producing once again substantial sediments with geological recordings.

As a result of renewed and vigorous weathering, calcium ions and bicarbonate would deluge the oceans. In what would be now warm ocean water, plentiful precipitation of calcium carbonate is understandable by the reduced solubility of calcite and aragonite at the higher temperature as discussed in Chapter 12. Warm-water limestone deposit layers would materialize directly above the glacier deposit strata since the transition from snowball to "sauna" was geologically rapid. Presence of these deposits accounts for them being labeled "cap carbonates" by experts in these studies. Enough carbon dioxide to precipitate 800,000 km^3 of material was presumably consumed.[†] Hypothetically, this is sufficient to cover the crust to a depth of 5 m. If accumulated in the air instead, it would amount to 0.12 atmospheres pressure. In fact, cap carbonates are thickest in regions where ^{13}C depletions in the carbonate imply higher temperatures. Layers within the cap carbonates are interpreted as having been accumulated at rates of about 0.4 m per year. (Compare that to the deposition rate estimated for the Cliffs of Dover that was a fraction of a millimeter per year.) The swing time between extremes, reflected in the border between observed ice rock sediment with its dropstones and the limestone capping it, might be as short as a few centuries. A long Arctic hiberna-tion was followed by a quick-thaw, wake-up call.

---

[*] B. Bodiselitsch, C. Koeberl, S. Master, and W. U. Reimold, "Estimating duration and intensity of Neoproterozoic snow-ball glaciations from Ir anomalies," *Science* 308, 239 (2005).

[†] P. F. Hoffman, A. J. Kaufman, G. P. Halverson, and D. P. Schrag, "A neoproterozoic snowball earth," *Science* 281, 1342–1346 (1998).

By the end of the Neoproterozoic, continuing landmass drifts had reassembled another supercontinent called *Gondwanaland* mostly in the southern hemisphere. (And recognizable to many as the contemporary jigsaw fit apparent between western South America and eastern Africa land outlines.) Geography's influence on carbon dioxide has changed again.

SURVIVORS

How could life persevere under such major long-term environmental disruption?

In the late 1970s, extreme-environment life forms were discovered. *Thermophiles* (heat lovers) were found near volcanic activity, in hot springs at volcanic rifts. Their counterparts, *psychrophiles* (cold lovers) were found in the frigid environment of Antarctica. It was already known that there are life forms that require neither light nor oxygen to exist. Consequently, the riddle as to survivors is not difficult to understand. They were primitive in structure though.

After the great snowball thaw, storms would not only resurrect the weathering processes on the re-exposed land, but also wash terrestrial nutrients into the seas where any photosynthetic bacteria that survived on the few thousands of perhaps small unfrozen pools present near volcanoes would bloom abundantly. Carbon dioxide levels in the atmosphere would drop sharply. Photosynthetically generated oxygen in the ocean would convert the vast amounts of accumulated ferrous iron, Fe^{2+}, into ferric, Fe^{3+}, which, being insoluble in that environment, would precipitate out profusely as the red bands of iron oxide stone plainly detected immediately above the thick glacial deposits in sediments and where the warm-water limestone layer was located.

Looking back at the puzzling findings, much can seem to have been addressed by the snowball Earth idea: all circumstantial but overwhelmingly consistent and convincing. Icebergs could still float around on the remaining liquid sea of a glacier-dominated surface. The encrustation of the planet with ice is instead viewed by some in a slightly less harsh version as "slushball* Earth." Describing the termination of these hypothetical cataclysmic glaciations remains controversial. An alternative rational interpretation of the evidence involves incomplete coverage by the ice layers. The idea includes the following descriptions. Global coverage with glacial drift rocks suggests that some open seas must have survived (to transport the material that will be dropped). Remaining uncovered landmasses surrounded by lowered sea levels were in permafrost condition allowing accumulation of vast quantities of methane from persisting life forms. When the glaciation began to reverse, the combination of increasing temperatures, violent weather, and rising sea levels would "suddenly" release the methane to the atmosphere. As an effective greenhouse gas, the methane would accelerate the temperature rise. Atmospheric methane gas gets oxidized to carbon dioxide relatively quickly and its effect on warming is substantial.

"Snowball" or "slushball"? Both remain speculative as no firm proxy has (yet) been found for thick ice in equatorial regions. In contrast, analysis[†] of rocks in Brazil at depths corresponding to the earlier of the snowball Earth incidents indicates the presence, not only of glacial deposits including dropstones, but also of shale rich in organic matter. Traces of compounds unique to microorganisms in the shale strongly suggest a diverse microbial community of photosynthetic character. That would be consistent with an ice-free domain or relatively thin ice covering. Whether this is just a local or short-termed indication has not yet been determined.

However it is looked at though, the idea is that some two to four cycles of icehouse-to-greenhouse swings occurred during the Precambrian Period from 750 to 590 Mya. After this, continuous tectonic motions presumably split the landmasses sufficiently apart and away from the equatorial zone, exposing more coastlines. In turn, such geographical changes make smaller landmasses subject to greater rain-induced runoff and weathering ending the snowball cycles. (For now?)

---

* Since the ice was presumably not slushy, some experts prefer the term *waterbelt* to slushball.
† A. N. Olcott et al., Biomarker evidence for photosynthesis during Neoproterozoic glaciation," *Science* 310, 471 (2005).

EMBEDDED COMPASSES

In reaching such bizarre conclusions as just presented, geologists were helped by additional clues. Magnetism embedded in the studied sedimented rocks indicated that those substances had come from equatorial regions. How? The direction in which magnetic substances point when formed (solidified) is toward the magnetic North Pole. Careful magnetic sleuthing therefore can reveal latitudinal origins of sedimentary formations. But this is conditioned on the requirement that no mechanical motions subsequently shift the position of the rocks. Among the useful and supporting observations was evidence at various depths in the rocks of North Pole/South Pole polar reversals, known to occur every few hundred thousand years. The magnetic properties in some deposits showed as many as seven reversals implying uninterrupted sedimentation times of perhaps a million years and lending credence to the necessary criterion that the magnetic directions had been locked into place, not shuffled by physical disturbances that would have made the deductions inconclusive. However, confidence in magnetic evidence assumes Earth's magnetic properties are well understood over the eons and that re-magnetization has not occurred.

CARBONATE ISOTOPE EFFECTS

Discovered in the layers just below the glacial ice rock debris, that is, in the older layers, was the carbonate substance called *oolite*, another limestone, a picture of which is shown here. These substances are inorganic pellets that grow by accumulating thicker and thicker shells.* Oolites are tropical water proxies. They are nearly spherical in shape, and physically quite delicate. If they had been moved from their original sedimentation settling sites, they would have fractured. Consequently, physical evidence, their shape, can indicate that they were not disturbed since settling. The highest oolites, closest to the bottom of the ice rock layer overhead, were formed when the snowball situation was evolving, that is, the period when the abundance of living things should have been undergoing considerable reduction by the plummeting global temperatures and from impending glaciation according to the snowball Earth picture just painted. Recalling from Chapter 14 on photosynthesis, bacteria near the ocean surface favor the use of the lighter isotope of carbon, leaving sea bicarbonate enriched in the heavier ^{13}C. However, since life had been decimated, it was anticipated that an isotopic measurement on the oolites would be relatively ^{13}C-poor. But the oolites were found to be ^{13}C-enriched, much to the initial dismay of snowball Earth's proponents. With a bit more insight, though, it was recognized that the carbonate from which the oolites were bioassembled at these times came from prevailing limestone sediments in the ocean since the traditional supply of fresh carbonate was reduced with the diminution of the chemical weathering process as glaciers spread. Oolite production utilized the carbonates from existing limestone that was dissolving in the sea water. That is, oolite carbonate skeletons' enrichment in ^{13}C was indeed consistent with the complete pre-snowball conditions.

---

* Even kidney stones are similar to oolites.

The ice-ball picture has most of life eliminated. Consequently, there aren't the usually abundant photosynthetic microorganisms fostering the isotope fractionation at this stage. Above the glacial ice rock, the ^{13}C in carbonate caps is depleted. But over millions of years, the $^{13}C/^{12}C$ ratio would have equilibrated to the ratio found in volcanic sources, as observed. With the passage of time after the snowball has thawed and warmed, at higher (younger) levels in the cap carbonate above the glacial rocks, ^{13}C enrichment again recovers, consistent with the re-emergence of biological activity that preferentially consumes the lighter carbon isotope. The behavior just described is strikingly illustrated in the figure here showing the relative ^{13}C depletion* from sites across the globe over this several hundred-million-year period. The shaded regions correspond to the alleged glaciations. Leading up to the glaciations (to the right of each shaded region) are significant depletions of the heavy carbon isotope. These are from the carbonate deposits below the glacial detritus. During glaciation itself, if/when the snowball effect pertains, there is minimal deposit of carbonates and so a gap appears in enrichment measurements since there is little in the way of useful samples. Enrichments in ocean waters during this stage should be maintained close to the volcanic value of -0.6% as shown. At the end of glaciation with warming and recovery of photosynthetic microbes, enrichment of the heavier carbon isotope in new carbonate deposits above the glacial rocks are the so-called "caps." The ages shown at the top of the figure are from uranium-lead radioisotopic dating. The vertical dark line at 716.5 Mya was from analysis of samples from the Franklin Large Igneous Province, a flood lava basaltic deposit from northwest Canada and into Greenland, covering more than a million km^2. Magnetic measurements of these deposits suggest that when they evolved the deposits were within $10°$ of the equator.

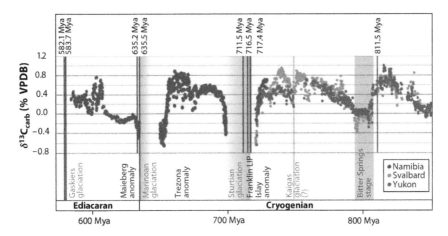

Some unresolved issues are associated with these measurements and their interpretation. The depletion before 811.5 million years (to the left of the line marker) ago does not seem to be associated with a glaciation. It has been tentatively attributed to a "true polar wander," a shift in Earth's orientation. The weak excursion evident at almost 750 Mya labeled "Kaigas glaciation (?)" may have been local (to the Kalahari in Africa) or not a glaciation at all. In general, if there is a glaciation, it appears there will be an accompanying negative enrichment excursion. But a negative enrichment excursion does not necessarily indicate a reciprocal glaciation has occurred.

Similar but weaker $^{13}C/^{12}C$ flip-flops are found in carbonates dated to other alleged mass extinction events (but of shorter duration than the snowball episode).

---

* Shown are percent departures from a standard: Vienna Pee Dee Belemite carbonate which has a ^{13}C enrichment of 0.011056.

SNOWBALL EARTH: BETA VERSION

Were there even earlier snowball Earth events? There are massive deposits of red iron oxide that date back to the Paleoproterozoic Epoch, that is, the early Precambrian about 2.2 billion years ago.

These truly ancient deposits mark the alleged onset of atmospheric oxygen buildup. One of the largest such deposits is in the Kalahari Desert region of southern Africa. The evidence is that the deposits have been relatively undisturbed over the ages.

Also present in that African location is a tremendous manganese deposit; some 4 billion tons of manganese (Mn) are estimated to have sedimented here. The age of these sediments is determined by radioactive dating techniques. The manganese deposits were found to be 2.2 billion years old. Of significance is that, like the iron oxide, the manganese deposit also originates in sub-oceanic sources. In the ionic form, Mn^{2+}, manganese is soluble. With oxygen present, the manganese could be oxidized to its insoluble Mn^{4+} version. Both Fe^{3+} and Mn^{4+} precipitate as oxides. The precipitating manganese, in sinking, eventually drops to ocean depths at which oxygen levels are minimal. Here, unsupplied with oxygen, Fe^{2+} is still sustainable. However, as long as there is Fe^{2+} present, the manganese oxide won't remain a precipitate because the ferrous iron has the ability to convert Mn^{4+} back to the soluble Mn^{2+} form according to the simple swap symbolized by $2\,Fe^{2+} + Mn^{4+} \longrightarrow 2\,Fe^{3+} + Mn^{2+}$. Iron oxide becomes the precipitate. Only when the iron sedimentation is essentially complete sequestering all the Fe can the manganese then deposit. The presence of the manganese deposit in the Kalahari field atop red iron bands helps support the idea that oxygen at this time (possibly supplied from *cyanobacteria**) was sufficient to first precipitate all the iron that had accumulated followed by precipitation of manganese oxide. The amount of carbon dioxide that would have to be photosynthetically consumed to release enough oxygen to account for the Fe/Mn sediments observed is estimated to be equivalent to 0.06 atmospheres, a bit less than half but still comparable to the maximum expected during the "more recent" Precambrian snowball demise of 590–750 Mya.

Manganese deposits are also found atop the glacial deposits associated with the more recent late Precambrian extreme temperature excursions previously discussed. The Kalahari site is unique though. It has ice rock present indicating a long-lasting snowball Earth event. In with the ice rock here, though, is a layer that is several hundred feet thick and that is flood basalt lava. Magnetic measurements on the solidified lava indicate that the deposit, with its identification as occurring during a snowball event, had been very near the equator at the time the lava hardened locking the magnetic direction into place. Note that the Kalahari today is at the Tropic of Capricorn, 23.5° (about 1,600 miles) south of the equator. Traces of long-lived radioactivities in the African deposits indicate an age of 2.2 billion years. Again, the concept of the snowball picture suggests that at that period the landmasses must have been assembled at the equator. The solar radiation level then was even lower than in the more recent Precambrian Period discussed initially despite higher CO_2 atmospheric levels, and the frozen world evolved…even in equatorial regions.

No evidence of any earlier snowball event than that of 2.2 billion years ago has been found. The indications are that Earth experienced very long-duration ice burials during the early Precambrian (Paleoproterozoic), and during the late Precambrian (Neoproterozoic), and at no other times. A repeat is regarded as unlikely, not only because the continents are dispersed away from the equator, but also because warming solar radiance has continued to increase as the sun ages.

Crucial to the development and appreciation of these ice scenarios has been carbon dioxide and its characteristic involvement in global descriptions.

---

* *Cyanobacteria* are a phylum of bacteria that obtain energy through photosynthesis but without *chloroplasts*, structures employed by more modern bacteria.

18 Food and Drink

DIGESTION AND CARBON DIOXIDE

We all know that the stomach secretes acid for digestion (and indigestion). The acid is hydrochloric acid, HCl, which is classified by chemists as a strong acid. In the stomach, it is referred to instead as *gastric acid* although the latter actually contains salts such as potassium chloride and sodium chloride as well. What is not realized by most people is the fundamental role that carbon dioxide plays in the physiological process of synthesizing hydrochloric acid in the stomach and that acid does not break down food.

We begin the exploration by setting the scene in "parietal" or "oxyntic" cells. These line the inner walls of the stomach. Oxyntic means acid-generating; parietal comes from the Latin word for "wall." The cells secrete hydrochloric acid at about pH 0.8, giving hydrogen ion concentrations about 3 million times greater than that found in blood. To obtain that concentration, about 1,500 calories are needed per liter of acid produced. Our main characters in this scheme are the following. Within these parietal cells are water, dissolved carbon dioxide, hydrogen ions (H$^+$), potassium ions (K$^+$), *carbonic anhydrase* (pictured here, a key enzyme in the process), and an enclosed organelle called the *mitochondrion* whose chore is to produce a substance abbreviated as ATP (*adenosine triphosphate*), the energy-producing molecule that we have seen before in discussing photosynthesis (Chapter 14) and respiration (Chapter 15).

Carbonic anhydrase

On the periphery of the parietal cell, facing away from the stomach contents, are three receptor sites which are chemically constructed so as to recognize – match up structural geometry docking regions with – and bind (receive) three different acid-triggering agents: (1) *histamine*,

Histamine

(2) *acetylcholine*, and (3) *gastrin*. The pharmaceutical products Pepcid AC, nizatidine (Axis AR), and ranitidine (Zantac) are designed to impede histamine.

$$CH_3$$
$$\overset{+|}{H_3C\text{-}N\text{-}CH_2\text{-}CH_2\text{-}O\text{-}C\text{-}CH_3}$$
$$\underset{CH_3}{|}$$

Acetylcholine

There are also large structures that sit in the wall of the parietal cell, extending from one side of the wall to the other that, in a sense, escort or channel specific chemical species into or out of the cell. One of these, on the side away from the stomach contents, exchanges the chloride ion (Cl^-) with the bicarbonate ion (HCO_3^-), swapping these two negatively charged species to maintain electric neutrality as is necessary. Three other such channels, facing the stomach contents, are responsible for exchanging H^+ with K^+ or with pumping K^+ or H^+ out of the cell and into the stomach. Prilosec OTC is a "proton pump inhibitor."

Gastrin is a polypeptide hormone, a chain of amino acids, whose linear sequence is abbreviated* as shown here and whose structure looks pretty much like the large molecule displayed.

Pro-Gly-Pro-Trp-Leu-Glu-Glu-Glu-Glu-Glu-Ala-Tyr-Gly-Trp-Met-Asp-Phe

Gastrin I (human)

The intricate sequence of digestive events now goes something like this. When the three receptors have been individually turned on by their dockings with histamine, acetylcholine, and gastrin, this produces the "go ahead" signal to the enzyme carbonic anhydrase inside the cell that next binds both water and carbon dioxide, nearly simultaneously, and hastens their otherwise slow rearrangement into and release as hydrogen ions and bicarbonate ions according to

$$H_2O + CO_2 \rightarrow H^+ + HCO_3^-$$

---

* Amino acids: Pro = proline, Gly = glycine, Trp = tryptophan, Leu = leucine, Glu = glutamic acid, Ala = alanine, Tyr = tyrosine, Met = methionine, Asp = aspartic acid, and Phe = phenylalanine.

The produced hydrogen ions migrate through the cell wall channel toward the stomach where they will be secreted. The bicarbonate reciprocally migrates away from the stomach side of the cell and will be transferred to extracellular fluid outside the stomach. But it is not quite that simple. A potassium channel effectuates importing potassium ions from the extracellular fluid outside the parietal cell in exchange for sodium ions, Na^+ from within. The hydrogen ions attach to one end of a H^+/K^+ exchanger, acting as a pump for transfer into the stomach as the necessary ingredient for any acid. That exchange is powered by the conversion of mitochondrial-released ATP to ADP (*adenosine diphosphate*), a transformation that provides the needed energy. That is, K^+ is brought into the cell by the pump in exchange for exiting H^+. At the opposite wall of the parietal cell, the bicarbonate ion finds the pump structure that will allow it to pass out into extracellular fluid in exchange for chloride coming into the cell. The imbalance of ionic charge (not detailed here) causes potassium ions and some sodium ions to diffuse out of the cell into the stomach. At essentially the same time back on the side of the cell abutting the stomach cavity, chloride ion inside the cell passes through its selective channel into the stomach. The overall effect is to use carbon dioxide in the presence of water to generate H^+ ions which are secreted into the stomach along with chloride ions, which were traded for the bicarbonate, providing ultimately the hydrochloric (gastric) acid used in digestion.

What the acid does is to cause the very large molecules, proteins, in the food to unravel. (Starches are mostly digested enzymatically in the small intestine.) But the acid does not itself break down the food into smaller molecular fragments. That is done by digestive enzymes such as *pepsin* which have their maximum efficiency in very acid environments.

When stomach acid exits the stomach and enters the intestine (the duodenum is the closest to the downstream part of the stomach), it triggers the release of *secretin*, a hormone that in turn causes the pancreas to secrete fluid containing large quantities of bicarbonate: about 9 g per liter, many times the level present in blood and producing a local pH of about 8, slightly alkaline and ideal for the behavior of pancreatic enzymes. What follows then is neutralization of the hydrochloric acid (which had reciprocally been produced using bicarbonate in the stomach) as

$$HCl + HCO_3^- \rightarrow Cl^- + H_2O + CO_2$$

The carbon dioxide is reabsorbed into the blood. As a pH 7.4 buffer, the blood ensures that the CO_2 is converted yet again into bicarbonate circulating with the rest of its kind (Chapter 10). Too much carbon dioxide can drop the pH (slightly) causing drowsiness, a common symptom following a meal. And on passing back to the lungs, the CO_2 from the bicarbonate is re-expelled as carbon dioxide gas. If this cycle of neutralization did not follow the stomach's acid production, ulcers would result from acid attack, most frequently in the duodenum.

Common over-the-counter remedies for "acid indigestion" take advantage of exactly the same chemistry just discussed. Two such items are shown here.

In Tums®, the active ingredient is calcium carbonate which follows the reaction

$$2\,HCl + CaCO_3 \rightarrow Ca^{2+} + 2\,Cl^- + H_2O + CO_2$$

Alka-Seltzer®, on the other hand, contains the combination of citric acid and sodium bicarbonate ($NaHCO_3$). When the tablet is placed in water, the famous effervescence that results is from the reaction of the citric acid with the bicarbonate to release carbon dioxide, essentially like in the equation for hydrochloric acid. It is not the bicarbonate in Alka-Seltzer that brings about the neutralization. When the bubbling is complete, sodium citrate remains in solution and this acts as a buffer to counteract the gastric acid of the stomach.

BREAD, BEER, CHAMPAGNE

Fermentation by microorganisms has been known for several thousand years. For most of the time, it was understood to be a process driven by living "things," even if they couldn't be seen. Bread, even in primitive form, has been around reportedly for 30,000 years. The use of yeast, microbes, and a type of fungus metabolize carbohydrates such as sugars and starches into carbon dioxide and alcohol. The carbon dioxide gas puffs up breads. Eduard Buchner, a chemist and botanist from Munich, studied the chemical processes of the contents of yeast, recognizing in 1897 that fermentation could be repeated synthetically, in the absence of yeast cells. For this, he was awarded the Nobel Prize for Chemistry in 1907.

In 1929, Arthur Harden from Manchester, England, and Hans von Euler-Chelpin of Augsburg, Bavaria, shared a related Nobel Prize for Chemistry for their studies on fermentation of sugars. Von Euler-Chelpin actually started out as an art student, but became fascinated by color. His interest in color led him to study science. He worked with a number of chemistry elite during his early years including Svante Arrhenius. In introducing the work of these two chemists on alcoholic fermentation, the following statement was made by the Nobel committee:

> Little more than a couple of centuries separate us from the time when men first began to perceive that the fermenting substance was sugar, which under the influence of a certain *something* was decomposed, with carbonic acid and ethyl alcohol as the final products of the decomposition.

In making Champagne (only from grapes grown in the Champagne region of France), the fermentation process converting grape sugars into alcohol and carbon dioxide traps the gas inside the bottle during its second fermentation stage. The process, including the critical second fermentation stage, was discovered only in the 1600s.

SODA

Champagnes, sparkling wines, and beer are saturated with carbon dioxide produced by the fermentation of sugar. Soda drinks are carbonated commercially. Thomas Henry, from Manchester, England, and father of William Henry of Henry's law (Chapter 10), seems to be the first to have produced carbonated water based on a design by Joseph Priestley in the 1770s. In 1794, Jacob Schweppe began selling artificially carbonated mineral water in Geneva, Switzerland, where he had originally been a watchmaker. In a couple of decades, flavors were added: ginger around 1820, lemon in the 1830s, tonic in 1858, and cola in 1886. Schweppes became an established brand in a major new industry. Nowadays, the carbon dioxide is supplied either as dry ice or as the liquid (at 1,200 pounds per square inch pressure: 80 atmospheres).

19 Fossil Fuels

We are all addicts of fossil fuels.

KURT VONNEGUT

Fuels are energy supplies. Carbohydrates in our diet are fuels. Enriched uranium is a nuclear fuel, a special category tapped by mankind starting about a half century ago. Wood is a fuel. Hydrogen fuels fusion in most stars. But distinguished from these are what are termed "fossil fuels": coal, petroleum, and natural gas. They are labeled with the adjective "fossil" because they accumulated eons ago, as geologists have believed for a long time. The term dates back to the mid-eighteenth century and originally referred to fuels dug from the ground (as opposed to wood, for example). Nowadays, usage is reasonably attached to the fact that these fuels have their origin in formerly living organisms. We currently consume fossil fuels at an estimated 1 million times the rate at which they were formed.

ORIGINS

The source of fossil fuels is ultimately photosynthetically driven, although there is a controversial alternative path we'll discuss later. Evolution of plant life on land presumably led to a burgeoning growth of species, especially of the "vascular" type – like trees with trunks – as far back as the Carboniferous Period more than 300 Mya. Leaves and even trunk-like material accumulated in vast quantities. Through chemical transformation while buried in environs from which oxygen was one way or another excluded, coal, petroleum, and natural gas formed. This is most convincingly argued by recognizing fossil remains in coal such as the fern leaves conspicuously visible in the figure here.

A French chemist and geologist, Alexandre Brongniart, suggested in 1810 that there must have been abundant vegetation during the "coal period," indicating a large proportion of carbonic acid in the air then.

Ferns, many of them gigantic, grew in or near swamps where they accumulated and, after expiring, would undergo anaerobic decomposition. Ultimately, the nature of the starting material (non-living organisms), the degree of lack of oxygen, and the temperature history combine to determine what ensues. Eight meters of peat yield 1 m of coal. A 2–4 m seam needs about 30,000–40,000 years to accumulate. Non-living organisms exposed to air on the other hand just return carbon dioxide to the atmosphere as the result of oxidative decay: respiration by microbes, for instance.

In thinking about coal, most know that there are different kinds: *lignite, subbituminous, bituminous,* and *anthracite*. But there is a relationship among these varieties. The same sequence gives the order of increasing hardness and decreasing success with which fossil remains can be found. A rational and fitting explanation is that their production involves, in the same order, increasing time, higher temperature, and higher pressure. Besides displays of fossils in coal, other evidence of coal's origin lies in the fact that deposits can be found layered between strata of sedimentary deposits.

There are both terrestrial and marine origins for fossil fuels. Some scientists feel that what distinguishes coal from oil is the choice between those two pathways. In either case, fuel production begins with settling or sedimentation of organic matter. It may be mixed with settling minerals as well. The average (an important qualifying adjective) content in sediments globally is 2% carbon. That amount varies between 0 and 100%, the latter being associated with coal deposits. Most carbon is in the form of carbonates though. Source beds, organic-rich sediments that undergo subsidence experience slowly rising temperatures. Chemical compositional changes, due to loss of hydrogen and of oxygen relative to carbon, occur slowly with time.

Initial mixtures of organic matter that ultimately give rise to oil and natural gas are very complex in nature, and referred to collectively as *kerogen*. Kerogen is organic matter, the altered remains of marine and other water-body microorganisms disseminated in sediments. The alteration involves degradation reactions at the bottom of a water column or at tops of sediment layers before their total burial. Kerogen is the most abundant organic carbon source there is. (Carbonates are not considered organic despite their organic origins.) Kerogen is a thousand times more plentiful than coal. Even in laboratory experiments, kerogen can be made to expel hydrogen and oxygen as water and more oxygen as carbon dioxide in a process called *diagenesis*. For temperatures between 60°C and 130°C, the usual result is to transform kerogen into oil. That is almost the definition of diagenesis,* the microbial and chemical transformation of organic matter at "low" temperature. Long times help brew the concoction. Millions to tens of millions of years can be required. But at temperatures that we would personally consider uncomfortably warm, yet which are low by geological standards (<40°C), even extremely long times won't break down kerogen.

The beginning of diagenesis, at depths greater than 1 km, is frequently (but not always) associated with the breaking of bonding interactions at the molecular level. This results in the formation of *bitumen*. The next step is to decompose the very large molecules – proteins and other polymeric structures – to smaller, more stable pieces. Controlled studies in laboratories have demonstrated that the presence of water during bitumen degradation increases yields and stability of hydrocarbon products. Carbon dioxide is also abundantly produced in the process. The lighter hydrocarbons and oils are buoyant – less dense – and the increased volume that results helps them separate from residual bitumen, migrating away and pooling. Water seems to aid in the separation process by its association with the bitumen which then becomes less miscible with the oil that is separating.

Diagenesis commences even in newly deposited sediments. Sediments from diatomaceous and microbial detritus have perhaps a couple of percent organic carbon. These sediments, though, deposit at rates of 2 m per millennium and can attain thicknesses of 500 m. If sediments undergo

---

* Diagenesis refers more collectively to transformations that occur in sediments. The Oxford English Dictionary has the following definition: "Transformation by dissolution and recombination of elements."

subduction as a result of which they experience higher temperatures, kerogen can break down to form oil and gas by *catagenesis*. This is a category of chemical transformation in which catalysis by minerals can be influential. The gas has large amounts of methane and CO_2. Liquid oil can pool and flow out of the sediments during diagenesis and tectonic motion. If the oil doesn't release at this stage, further subduction and consequential temperature increases lead to cracking of the hydrocarbons to form natural gas and other light molecules. Catagenesis is generally regarded as the process mainly responsible for formation of oil and gas. Time and temperature are the empowering factors. Composition of the kerogen and the structural properties of the source beds are also very influential. In recent years, an additional role of water and minerals has been recognized. These indirect influences serve as reactants, catalysts, and as controllers of porosity and permeability of source beds. Many of these ancillary influences are newly conceived and remain controversial.

If diagenesis were to continue for long periods at temperatures between 100°C and 200°C, natural gas – mostly methane (CH_4) – would result. Interesting compositional trends imply something more detailed about origin. In fact, the persistence of the hydrocarbons (compounds containing only hydrogen and carbon) ethane (C_2H_6), propane (C_3H_8), and butane (C_4H_{10}) in laboratory studies despite the temperatures used and the amount of time allowed for reaction suggests that natural gas production mechanisms are far from simple. For fuels containing concentrations of hydrocarbons with odd numbers of carbons in the range C_{15} to C_{21}, the interpretation is that these were made from marine plants. For C_{27} to C_{35}, the source is terrestrial plants. Carbohydrates and *lignins* require more extensive "brewing" for conversion into hydrocarbons and usually end up as coal, anthracite being the final product. Coal is 65–95% carbon and ranges in age from 30 to 300 million years. Anthracite is nearly 100% carbon with no traces of vegetation remaining.

As noted, the porosity of minerals in which the oil sources are located depends on the carbon dioxide produced during diagenesis and also on the acid pH that results. These chemically weather (corrode) pores, widening them and consequently facilitating the migration and pooling of oils. In trapped oil deposits, 80% of the pores are filled with oil and the remainder with water. The presence of water in contact with oil and at the correct temperature leads to microbial biodegradation of the oil brought about by the presence of deep, subsurface bacteria. More than half of the planet's oil is biodegraded heavy oil or tar sand deposits. The largest accumulation is *not* under the Arabian Peninsula but in a Venezuelan belt (nearly 200 billion m^3). The third largest deposit is in western Canada, the Athabasca tar sands. Under anaerobic (oxygen free) conditions, so-called methanogen bacteria* can convert CO_2 to methane. This process is also favored by high pressures. However, for this to be viable, relatively common chemical sulfates must not be very prevalent because of their toxicity to the microbes. Sulfate is SO_4^{2-}.

Deep reservoirs (down to 4 km) of petroleum often are associated with high temperatures since temperature increases a couple of degrees centigrade for every 100 m in depth. There is rather general agreement that biodegradation in deep reservoirs is by anaerobic microbial metabolism. However, despite a lot of searching, these performers have not yet been identified. As the temperature of the deep reservoirs exceeds 80°C and approaches 120°C, biodegradation by the bacterial community becomes more difficult and eventually ceases. Living species can't exist at higher temperatures. (Think pasteurization.)

In commercial processing of petroleum, higher temperatures are required to "crack" the mixtures, breaking them down into useful derivatives such as gasoline (octane). In nature, cracking of oils in sediments at lower temperatures is explained by the long times involved and by the probable role of co-mingled clays acting as catalysts. Bacteria might play a role here as well.

There are also petroleums generated hydrothermally at sea vents. The oil and gas brewed there arises via hot circulating water, between 100°C and 300°C, depending on location, and at pressures up to 200 atmospheres. The action is actually more rapid than the diagenesis in sediments undergoing subduction that we just discussed.

---

* Actually not bacteria, but an even more primitive kingdom of single-celled microorganisms referred to as *archaea*.

ALTERNATIVE SOURCE

Not everyone is convinced that diagenesis is the origin or lone origin of petroleum and natural gas. Thomas Gold, a geologist from Cornell University, has argued for years* that the hydrocarbons are naturally occurring substances with an inorganic (as opposed to organic) derivation. The "deep Earth gas theory," briefly, is that hydrocarbons formed through inorganic processes at mantle-like depths and then migrated toward the surface.

The most recent activity with this view emerged from a joint American–Russian effort in 2002 to demonstrate, successfully, that petroleum products could be made from mineral carbonates, water, and iron oxide at high temperature (1,500°C) and high pressure (50,000 atmospheres). Geologists agree the chemistry is possible, but insist that most commercial petroleum is organic and cite "bio-markers," unusual hydrocarbon signatures peculiar to marine plants or terrestrial plants. The inor-ganic-route advocates note that these could simply be contaminants in inorganic hydrocarbons. Further evidence in support of the inorganic process is the abundance of methane – the lightest, most abundant component of natural gas – in volcanic ocean vents where there are insignificant amounts of biological sediment. Also noted is the ubiquitous presence in oil of a class of compounds called *porphyrins* frequently associated with a metal constituent. Porphyrins are abundant in plants. The metals found with porphyrins in oil, though, are nickel and vanadium, two metals rarely associ-ated with porphyrins from living matter. The latter are invariably magnesium (as in chlorophylls) or iron (as in hemes). Yet magnesium or iron porphyrins aren't found in petroleum. Contamination is always an important issue, of course.

Finally, analysis of the hundreds of variations in structure of hydrocarbons in petroleum from a collection of oil fields around the world indicates they have a common origin with formation tem-peratures of 700–1000°C and very high pressure, conditions corresponding to depths of the upper mantle and way too hot to be ascribable to a sedimentary, organic origin.

BUT WAIT! THERE'S MIRE

Mire is another word frequently used for peatland. Other terms are moors, bogs, fens, muskegs, and pocosins. Peat is fuel used in many parts of the world, found as far north as the Arctic and down to the tropics. Deposits average about 2 m in thickness. Some 30–70 tons of carbon as peat are typically deposited per year per km². Accumulations as much as 10 m are common. Peat covers 4 million km² of land, about 2–3% of Earth's terrestrial area. The largest source is in Siberia and covers an area that is twice that of all of Sweden. Peat is the accrual of decayed or decaying veg-etation, the most common of which is slowly decaying sphagnum moss. The deposits accumulate over thousands of years in an environment that must be a shallow wetland and preferably one that might actually be subsiding, allowing flooding waters to continually block the access of oxygen and thereby slow down the ordinary decomposing action that would release carbon dioxide back to the atmosphere. A peat bog is pictured here. The organic material is slowly degraded by the action of bacteria and fungi in processes that are not yet completely understood and that last tens of thousands of years. The process is called *peatification*.

---

* Thomas Gold and Freeman Dyson, *The Deep Hot Biosphere: The Myth of Fossil Fuels*, Copernicus Books (2001).

Peatification is slow, showing accumulation rates of about 1 mm per year. Presumably because there is replacement of peat, even though it is sluggish, the International Panel on Climate Control continues not to consider peat as a fossil fuel, the latter being restricted to coal, oil, and gas. But considering the age of the peat deposits and the rate at which they are combusted by natural or man-made processes *versus* their regrowth, this does seem a curious mis-categorization.

Peatification over geological time periods and under the proper conditions converts peat into coal. That slow process, one that requires heat, has been referred to as *coalification*. Estimates are that peatification has been going on for nearly 400 million years with a total global accumulation equivalent to 2,000 Gt of carbon dioxide. That is comparable to the total atmospheric content of 3,200 Gt. It is also similar to the content of coal reserves which has been reported recently by the IPCC as about 500 Gt carbon and by the U.S. Energy Information Agency as about 1,000 Gt corresponding, respectively, to 1,800 and 3,600 Gt carbon dioxide.

If or when the peat deposits dry, as with lowered water levels brought about naturally or by man, they are subject to incineration both in the form of extensive fires (see Chapter 9) and hidden smoldering combustion, back into carbon dioxide. Nowadays, drained peatlands are releasing about 2 Gt of carbon dioxide annually.

20 Isotope Stories

Accurate determination of the ^{18}O content of carbonate rocks could be used to determine the temperature at which they were formed.

H. C. UREY, 1941

STORYTELLERS

A variety of techniques are available for tracking the history of carbon dioxide through the ages. Working backward from the present, these include:

- Instrumental records (like thermometers) over the past half century
- Tree rings going as far back as several millennia
- Coral reefs whose contents have information dating back some hundreds of centuries
- Ice cores whose contents have been traced back to nearly a million years recently*
- Lake sediments for millions of years of data
- Deep sea sediments now analyzed as far back as nearly 200 million years
- Land sediments can be found as old as billions of years

Data registered in these depositories will be looked at briefly and some comment on assessing the ages of the records will also necessarily be included. We will look first at the facts and then at some of the narratives woven from them.

EARLY ICE AGES, BRIEFLY

The meaning of the opening quote will be made clear. Its author, Harold Urey, is regarded as the father of the field of *paleotemperature* studies, learning about ancient temperatures. But among even earlier attempts were some that sought to explain intriguing observations, which is what science is about. In particular, the idea of one or more "ice ages" began to evolve nearly two centuries ago. (Recall the fascinating event chain involving prolonged ice ages mentioned in Chapter 17.) A Swiss mountaineer, Jean-Pierre Perraudin, became intrigued by the granite rocks he would encounter high in the Alps, some as large as houses. Their appearance was incompatible with minerals in the region. Their mass and size suggested that being transported there by water was not a likely explanation. And then there were the parallel scratch marks on rock surfaces, shown here as an example. In 1815, he consulted with a Mr. Charpentier with his views. As a result, Jean de Charpentier, a Swiss-German geologist, was the first to systematize, by the 1830s, the gathering evidence favoring an ice age, despite his initial skepticism.

---

* And will be extended back to 1.5 million years shortly.

Charpentier led a personal tour of the Alps accompanied by the well-known fossil expert, Louis Agassiz (shown here), in 1836.

Agassiz went along, expecting to confirm his own doubts about Charpentier's ideas and furthermore, anticipating he would be able to point out the fallacies in the ice age hypothesis. Instead, Agassiz became a convert. That year, he presented his new views at a science gathering in Switzerland and the presentation was greeted with rage. In his book *Etudes sur les Glaciers* of that year, Agassiz wrote

> The perched boulders which are found in the Alpine valleys…occupy at times positions so extraordinary that they excited in a high degree the curiosity of those who see them. For instance, when one sees an angular stone perched upon the top of an isolated pyramid, or resting in some steep locality, the first inquiry of the mind is when and how have these stones been placed in such positions, where the least shock would seem to turn them over?

The mysteriously misplaced, alien stones like that shown here had been known for some time and were originally given the name "erratics" by de Saussure in 1779, a person we previously encountered in discussing the early history of photosynthesis research (Chapter 14). The stones are still known as erratic boulders. The prevailing consensus at the time was that these were carried by water flow, perhaps on ice floats.

Charles Darwin, in a short 1839 note about Antarctic discoveries, published in the Journal of the Royal Geographical Society of London an observation of a "Rock seen on an Iceberg." Published with the note was a sketch, reproduced here, of what was observed.

526 Mr. JOHN BALLENY's *Antarctic Discoveries in* 1839. [March,

¼ of a mile of an iceberg about 300 feet high, with a block of rock attached to it, as represented in the following woodcut from a drawing made on the spot by Mr. John M'Nab, 2nd mate of the schooner.

He describes the rock as a block of about 12 feet in height, and about one-third up the berg: it is unnecessary here to make any observation upon this very remarkable fact, as Mr. Charles Darwin has appended a note to these extracts, pointing out the value of such an evidence of the transporting power of ice:* we will, therefore, only add that this iceberg was distant 1400 miles from the nearest *certainly-known* land, namely, Enderby's Land, which

In 1842, Joseph Adhémar published *Revolutions of the Sea*, a quantitative argument for the ice ages being caused by motion of the sun and moon following a 22,000 year cycle, an idea greeted with enormous skepticism at the time, an almost obligatory greeting for many a revolutionary scientific idea.

James Croll, an Australian, self-educated school dropout, migrated from job to job eventually settling down as a janitor. This gave him time to pursue reading and expanding his interest in science. Croll was inspired by the idea of glacial periods and Adhémar's theory. Croll, the janitor, worked out the astronomical mathematics and physics by himself and in 1875, published *Climate and Time*, which became very influential in changing thought about climate influences. Croll linked climate directly to small variations in Earth's orbit about the sun. Even Charles Darwin, in his 1859 *Origin of Species*, mentions.

erratic boulders and scored rocks plainly reveal a former colder period

Ever since Louis Agassiz's postulation in 1840 of an "ice age" sometime in the Earth's past, scientists have questioned climate variations that might have occurred over the eons. But for the next century, there was no *quantitative* technique that could verify any effect. There existed explanations and there was growing qualitative evidence, such as scratched rock surfaces and enigmatic placings of boulders, but actual data were lacking. How this situation evolved and the relationship of the developments to carbon dioxide follows.

ISOTOPE EFFECTS

The chemistry of a substance depends on the behavior of the outermost (so-called valence) electrons. These, in turn, are influenced by the positive nuclear charge holding them in place. How heavy or massive the constituent atoms are does not enter into this consideration – at least to a very good approximation. And mass is what distinguishes isotopes of the same chemical element (see Chapter 1). However, mass differences among isotopes of the same element do moderate minute variations in behavior, especially in the lighter elements. Mass differences may be taken advantage of for a variety of purposes. The fissionable ^{235}U differs in mass from the abundant, non-fissile ^{238}U by about 1%. For carbon dioxide, the ^{13}C–^{12}C difference is more than 8% and the ^{18}O–^{16}O difference is more than 12%.

The practical use of the minor variations associated with different isotopes in chemical systems was first recognized by the American chemist (Nobel laureate for the discovery of *deuterium*, the isotope of heavy hydrogen) Harold Urey in the 1940s. First getting a degree in zoology, Urey went on to graduate studies in chemistry, and then on to academia. Much of his work on isotope chemistry was done at Columbia University. Urey was the first to explain the basis of chemical *isotope effects*. As an example that we'll also revisit later, consider the carbon isotopic composition of living tissue. The distribution of the carbon isotopes depends on whether, for instance, carbon dioxide is from the atmosphere or is from that dissolved in water; on the physical and chemical processes for fixing the element in some final molecule; and on secondary processes such as decomposition of decaying organic matter by bacteria. Urey suggested that the isotopic components of calcite that has been secreted by microorganisms might yield information concerning ocean temperatures and conceived of the idea of the isotope thermometer and its application to geochemical systems. Urey showed that the partitioning or fractionation of the isotopes of oxygen between water and carbonate – how they appear in different ratios in each – depends on temperature. This was the start of the field that is known as *paleoclimatology*, the study of ancient climate, a term dating back at least to 1920. Paleoclimatology involves uncovering temperature records.

WHY ANY EFFECT?

A naive picture of one of the factors that contributes to how easily a chemical reaction might proceed would be the ease or difficulty of breaking a chemical bond. That takes energy. Ordinarily, any two atoms bound together are oscillating back and forth slightly about some average separation distance. Adding energy can cause the atoms to vibrate more and more vigorously, spreading them to greater and greater separation extremes during their vibrations, until eventually, with enough energy, the bond would "pop." At first, the vibrational frequencies are minimal (although still very high from our perspective – some billion trillion oscillations per second). The heavier the isotopes involved in a bond, the slower, more sluggish, is that vibration. Light isotopes vibrate faster. To rupture a chemical bond, two linked light isotopes thus have a head start over a pair of heavy isotopes if one considers the energy input needed to eventually rupture the bond so some chemical reaction can take place. The difference is quite small, but has measurable consequences. A useful generalization is that in comparing two bonding situations, the one involving the heavier isotope preferentially survives rupture. Furthermore, this advantage is most prevalent at low temperatures.

In addition to the effects on chemical bonds, there is also the speed (velocity) with which a molecule moves around. Speed is determined by the temperature (hotter means faster) and by the weight of the molecule moving. For gases at the same temperature, all the molecules have the same energy, on the average, associated with their motions through their surroundings. This energy is referred to as kinetic energy or translational energy. But since the kinetic energy of a body is proportional to mass and also to the square of the velocity, two molecules with different masses will have different velocities, even though their energies could be absolutely identical.* At the same temperature, a heavy molecule moves more slowly than a lighter molecule. In the case of CO_2 exchanging between air and the ocean's surface, this could be manifested as a slight difference between the amounts of light carbon dioxide and heavy carbon dioxide in the atmosphere above the ocean, for example. Variations in isotopic abundances – the fractions of each present – in any molecular species are referred to as isotope *fractionations*.

---

* For example, carbon dioxide gas at room temperature has an average speed of 378.6 m per second (nearly 850 miles per hour) for the dominant isotopic composition with ^{12}C and ^{16}O. But if the slightly heavier molecule having ^{13}C were considered, its speed would be 374.4 m per second.

ISOTOPE FRACTIONATIONS IN NATURE

In the case of evaporation (escape from a liquid surface into the atmosphere), using water (H_2O) as an example, it is understandable that ^{16}O, the lightest (fastest) stable isotope of oxygen, would be concentrated (ever so slightly) in the vapors. This slightly faster H_2O with ^{16}O escapes the liquid easily, (it is not only faster but its bonding to the rest of the liquid is more easily disrupted) leaving complementary enrichment of heavier isotopes remaining in the liquid. Water vapor in the atmosphere is enriched in ^{16}O and depleted in ^{18}O. Ocean water has the opposite situation. Measurements confirm this but they also indicate what looks like a complete contradiction. Typical analyses of *fresh* water actually have negative enrichments – depletion – of the heavier oxygen isotope ^{18}O. Why? If the science of isotope fractionations is to be acceptable, such critical questions and/or challenges must be dealt with. The ostensible disagreement is quite understandable because such fresh water originates *via* rain or snow, mostly from water vapor that has evaporated from the sea. And that, as we just explained, has more of the lighter isotope. Interesting, *n'est-ce pas?*

^{16}O as part of the water molecule evaporates readily from tropical ocean waters. The water vapors (humidity) travel, sometimes over extreme distances before precipitation reverses the process. Precipitation as rain preferentially condenses those water molecules with the heavier, more tightly bonded ^{18}O in the water molecule. Any water vapor remaining as humidity is consequently further enriched in the lighter ^{16}O. As the air masses move further toward the poles and inland, temperature drops allow additional precipitation as rain and, eventually, snow. By the time the polar regions are reached, although the water vapor content has been reduced, the very cold region still produces snow fall. Now the water molecules comprising the snow are "very enriched" in the lighter ^{16}O.

For comparison to oxygen in water, oxygen as O_2 in the air follows a very different cycle and has a relative ^{18}O enrichment equal to +2.35%. That for oxygen in carbon dioxide in the air is +4.1%. The enrichment of heavier isotopes for both atmospheric oxygen and atmospheric carbon dioxide suggests they are not predominantly supplied as a result of evaporation from the oceans.

MASS MEASUREMENTS

Very precise determinations of slight changes in the abundance of measured isotope fractions are what enable researchers, starting with Urey and his group, to take advantage of the history record in layered depositories. The instruments responsible for this are called mass spectrometers, a diagram of which is shown here.

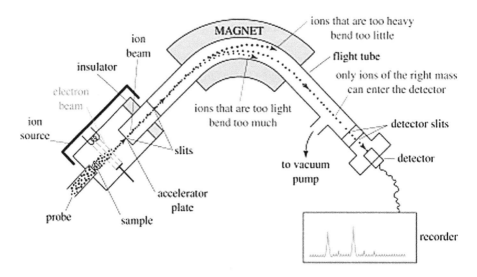

A brief explanation of how the device works goes like this. A small sample, perhaps of carbon dioxide, is fed into the device (at the left in the diagram) and converted to positive ions when struck by fast electrons which serve to knock out other electrons from the gas molecules. The ions are accelerated to a common speed by electrically charged plates forming a beam that travels between the poles of a magnet. It is well known from physics that a charged particle moving through a magnetic field experiences a force that bends the path of the ion into a circular route. But, just as a wind force can turn a light weight sailboat more effectively than it can turn a heavy sailboat, the magnetic force deflects the lighter ions more than the heavier ions. After traveling a fraction of a circle's orbit, the ions exit the magnetic field and continue subsequently in straight trajectories until being recorded in a detector. The number of ions of different masses is registered at easily distinguished deflection amounts indicative of their masses. A schematic of a detector's record is illustrated at the bottom right of the diagram here.

The problem that arises is that investigators are looking for very slight changes in the fraction of ^{13}C or of ^{18}O compared to those fractions in a reference standard. What is done to expose the extremely tiny fractionation differences is to switch back and forth between a sample and a standard, comparing the two very carefully. This is somewhat akin to trying to take two nearly equal weight children and determining their weight difference to a fraction of an ounce when their total weights might be a thousand ounces (about 60 pounds). A precision seesaw could be used so that the two weights could be counterbalanced very closely and their differences determined from where they were exactly seated. As the "counterbalance" for isotope ratio mass spectrometry, readily available standards with constant isotope composition are employed.

Enrichments and depletions of one isotope *versus* another of the same element are usually expressed in terms of a deviation from an agreed-upon standard (*std* subscript in the expressions here). The "standard" for oxygen is the isotopic composition of "mean ocean water." Likewise, the international standard used for carbon isotopes is Pee Dee Belemnite.* The comparison is conventionally reported as follows[†]

$$\frac{(^{18}O\,/\,^{16}O) - (^{18}O\,/\,^{16}O)_{std}}{(^{18}O\,/\,^{16}O)_{std}} \equiv \delta\,(\text{delta})$$

where $^{18}O/^{16}O$ symbolizes the ratio of fractions or abundances of the ^{18}O and ^{16}O isotopes in the substance and delta is the symbol representing what will be a very small relative difference value. For fresh water, a relative depletion δ of ^{18}O can amount to as much as -6.0% relative to the ocean (*std*) as is found deep inland in the polar regions. Keep in mind, though, that oxygen is only about 0.200% ^{18}O to begin with.[‡] This -6.0% departure would reduce the isotopic abundance of the heavy oxygen isotope to 0.200% · (1.00−0.06)=0.188%, a small absolute change indeed, but one that is quite precisely measurable by modern instrumental techniques. The snow in Antarctica's interior has the most isotopically fractionated oxygen found in nature. If all the current ice sheets were to melt, the ocean's ^{18}O fractionation, currently zero by definition (it's the "standard"), would drop 0.126% owing to the contribution of the new fresh water previously enriched in the light isotope. Ice sheets comprise about one-50th of total global surface waters. Following the same logic, if an ice age were to increase glacial mass, dropping sea level, the ocean's ^{18}O fraction would increase accordingly. Not surprisingly then, ^{18}O fractionation can serve as a proxy for ice/glacier volumes over time as mentioned in Chapter 8.

---

* Belemnite is a cretaceous marine fossil carbonate, found in large quantities in South Carolina, whose international standardization accepts the value for $^{13}C/^{12}C$ abundance ratios, for example, as 0.00112372.

† Usually expressed in parts per thousand, but we will use percent because it is much more familiar to most readers.

‡ Oxygen consists of 99.762% ^{16}O, 0.038% ^{17}O, and 0.200% ^{18}O. See Chapter 1.

CARBON ISOTOPE FRACTIONATIONS

The origin of natural fractionations is both interesting and important. In the previous sections we discussed physical enrichment paths for oxygen. Observed chemical enrichments in oxygen are mostly ascribable to oxygen's production through photosynthesis. In that quite complex process, the oxygen comes from the water that is involved in the synthesis of sugar from carbon dioxide and water in the leaves. The water, absorbed as liquid by root systems for example, is enriched in lighter isotopes. But photosynthesis occurs in the green leaves where, among other things, evaporation losses of water deplete those lighter isotopes. In contrast, during fixation of carbon dioxide, both physical and enzymatic processes favor the lighter isotope of carbon relative to carbon dioxide. The explanation is that the bonds between light isotopes are easier to break owing to the leg-up in carbon–oxygen bond stretching on the way to reaction. In marine plants and algae, carbon dioxide fixation via the Calvin photosynthetic pathway results in fractionation values

$$\delta \equiv \frac{(^{13}C / {}^{12}C) - (^{13}C / {}^{12}C)_{std}}{(^{13}C / {}^{12}C)_{std}}$$

of about –1.5 to –2.2%. Negative values imply relative loss of the ^{13}C isotope being compared to the standard ^{12}C. The carbon isotope standard is the fossil limestone Pee Dee Belemnite as mentioned before. Keep in mind that for every individual step a molecular synthesis process will have its own isotope effect that could be positive or negative, large or small. The cumulative isotopic discrimination is the percent quoted.

The combined processes leading to the incorporation of carbon dioxide into glucose by photosynthesis according to the Calvin cycle (the particular "C_3" cycle followed by the great majority of all plants and detailed in Chapter 14 on photosynthesis) leads to a measured and mostly understandable (at least to experts) isotope fractionation of –2 to –3%, a *depletion of the heavier carbon isotope in terrestrial plants.* Fossil fuels such as coal and peat are derived from terrestrial plant material and have a relative ^{13}C fractionation of about –2.5%. The small minority of plants that follow the known alternative photosynthesis pathway – a "C_4" cycle used by sugar cane, hot-region grasses, corn, and other crops – has about half this degree of isotope fractionation.

Unlike familiar plants, many algae found in the sea, instead of utilizing carbon dioxide, use the more abundant dissolved bicarbonate ion, HCO_3^-, for their photosynthetic feed material. It so happens that the chemical equilibrium between carbon dioxide and bicarbonate itself is known to have a fractionation of carbon isotopes to the extent of about +0.8%, that is, favoring the heavier isotope in bicarbonate. Accordingly, the heavier carbon is then incorporated into the photosynthesis cycle and the resulting isotopic discrimination is lessened compared to that for terrestrial plants. Photosynthetically derived carbon reaching the seafloor has a relative ^{13}C fractionation of about –1.8% compared to terrestrial fossil values around –2.5%. Tubeworms at deep sea vents have values ranging between –1.2% and –0.3% supporting the understanding that their carbon sources are not from photosynthetically derived carbon.

Aquatic plants also have a foraging capability in that they are able to biologically compensate for situations where the amount of available carbon dioxide or bicarbonate is low. There exists, in effect, a means of pumping dissolved carbon dioxide and bicarbonate (collectively called dissolved inorganic carbon) into cells. In the absence of enough "feedstock" to be supplied passively to the cells directly by diffusion, these pumping mechanisms are turned on. Since both carbon dioxide and bicarbonate are assimilated following pumping, the conventional isotopic differences between them are considerably washed out. Isotope studies should reveal periods of low carbon dioxide levels in equilibrium with the ocean using specimens of the appropriate species.

T RECS

It shouldn't be at all surprising to learn that isotope fractionation is affected by temperature. This is what Nobel laureate Urey originally found in his theoretical exploration of equilibrium involving different isotope compositions. For instance, the fractionations observed in phytoplankton that grow in cold water are higher than for those in species found in warm water. As soon as Urey, always the complete scientist, realized the possibility of isotope abundances serving as a recorder system – a proxy – for temperatures, he and his students set out to confirm their hypothesis. By the early 1950s they had determined the first calibration relating temperature to oxygen isotope fractionation in carbonates. They did this using the ocean as their laboratory and by analyzing the shells of marine organisms grown in seawater of known temperature. Their results illustrate the very strong correlation between temperature during growth and the percent difference of ^{18}O content in carbonate relative to a reference sample. Warm environments (toward the upper left) have reductions in heavier isotope components. So we see, isotopes are telling a story...and can serve as proxies.

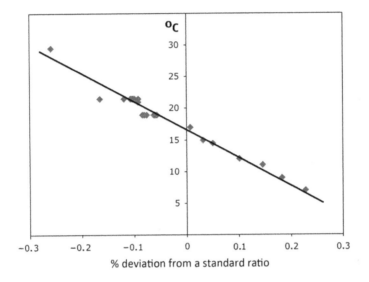

Similarly, samples of carbonate removed from successive growth layers of single marine shells that had grown over a year in a location off the coast of California with a known record of sea temperature were analyzed by Urey's collaborators for their $^{18}O/^{16}O$ ratios. Converting these ratios using the previously pictured calibration illustrated very well the consistency of the oxygen isotope "recording" of winter–summer temperature variations as diagrammed here. The smooth curve is just to guide the eye between the high temperatures of 21°C and the lows of 15°C of summer and winter, respectively. The researchers did recognize that their interpretation of the observed behavior depended also on possible variations in isotopic composition of seawater. But they regarded those concerns as ultimately of minimal influence.

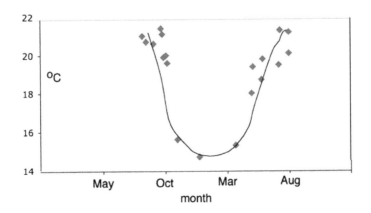

In 1967, Nicholas Shackleton* of Cambridge University analyzed core samples of foraminifera in deep sediments where the temperature environment of the foraminifera lives near sea bottoms is constant. He found ancient isotope fractionations paralleling those found in shallow ocean dwelling plankton. The result fostered the interpretation that the observed variations in each were due to changes in the ocean isotopic composition owing to glacial *versus* interglacial isotope effects. In fact, from these variations, ice volumes and sea levels could be inferred by analysis of buried marine sediments. An annoyance with these studies is the requirement that the foraminifera samples be well-shielded, else dissolution and re-precipitation could have altered isotopic ratios from their native values. More detailed analysis of complications may be found in volume 8 of the *Treatise on Geochemistry, 2nd edition: Elemental and Isotopic Proxies of Past Ocean Temperatures.*

We have looked at some of the short stories enabled by isotope measurements. It is time to move on to novellas.

MORE COMPLEX INFLUENCES ON ISOTOPE FRACTIONATION

Carbon dioxide, photosynthetically incorporated as carbon into living marine microorganisms, contributes to a steady shower of organic matter down to great depths. But the organic matter, having been enriched by photosynthesis in the lighter carbon isotope, gets decomposed into dissolved inorganic carbon by bacterial action. Deep saline waters are thus fed ^{12}C from the warm surface waters by biological activity. Biological activity should reflect the organism population distribution in the ocean environment. The global distribution implied for active phytoplankton is shown here. Coastlines, most especially in the northern hemisphere, show the largest surface concentrations of phytoplankton. Population densities translate into productivity regions.

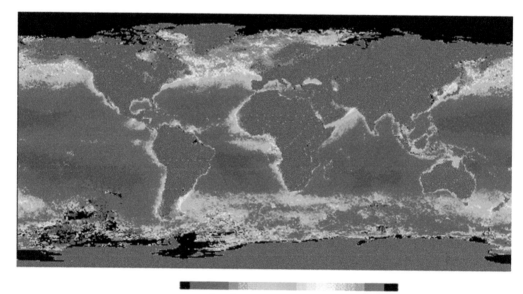

.1 .2   .4   .6 .8 1   10
Phytoplankton pigment concentration (mg/m3)

The effect of productivity we are now exploring is illustrated in the double graph here for water sampled in the North Atlantic. The graph on the left shows a depth profile of dissolved inorganic carbon.[†] The dissolved inorganic carbon is removed by photosynthetic activity in surface waters.

---

* N. Shackleton is a distant relative of the Antarctic explorer, Ernest Shackleton.
[†] The contributions to ΣCO_2 have been converted to their equivalent amounts of CO_2 as grams per kilogram of seawater.

The lowest amounts of dissolved inorganic carbon are at zero depth where sun activation of photosynthesis is easiest. On the right is a corresponding depth profile of ^{13}C relative isotope fractionation in dissolved inorganic carbon (in solution, not photosynthetically incorporated into organic molecules) showing maximum enrichment (of ^{13}C) at the surface. This is a reasonable interpretation since the maximum photosynthetic activity is at the surface, removing ^{12}C preferentially, leaving seawater there enhanced in the heavier carbon-13 isotope even though the amount of CO_2 is minimal. Photosynthesis exhibits an average ^{13}C isotope effect of around −2%. The residue of unutilized, dissolved inorganic carbon dioxide has consequently lost more of its lighter isotope and is enriched in ^{13}C especially at the ocean's surface.

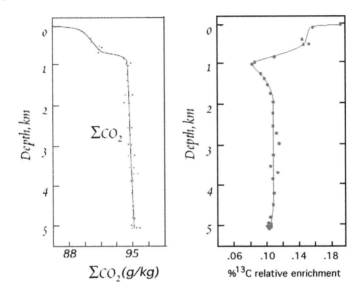

The longer a (large) sample of water remains at the surface where photosynthetic activity is ongoing, the greater is the accumulated discrimination against the heavier ^{13}C isotope into organic biomass with the reciprocal enrichment of that heavier isotope in the dissolved inorganic carbonates.

Even *deep* ocean waters (1,000 m and deeper) in the North Atlantic remain slightly enriched in ^{13}C. The values are around 0.10% shown in the right figure. Those waters had only relatively recently been near the surface (originating near Iceland in the so-called Denmark Strait). In contrast, deep waters in the Pacific Ocean come from near Antarctica and are much older because travel time *via* ocean currents to the Pacific is centuries long. Oxidation of descending organic material previously enriched in the lighter isotope of carbon by ^{12}C-favoring photosynthesis lowers the deep sea's fraction of ^{13}C as the water "ages." Consequently its ^{13}C is lower (the lighter ^{12}C is higher) than for the North Atlantic. Although this description has been over-simplified here, it implies that the carbon isotope composition of carbon dioxide derivatives can be used to roughly map out otherwise unseen ocean currents that have functioned over the eons below the surface. It is pretty much accepted that the deep ocean waters do not re-surface for roughly a millennium.

"YOU ARE WHAT YOU EAT"

In 1978, DeNiro and Epstein* explored how diet of animals affected their tissue isotopic composition. In their results, illustrated in the figure here, the straight slanted line indicates what would have been expected if there were no further fractionation of carbon isotopes by animals relative to their

---

* M. J. DeNiro and S. Epstein, "Influence of diet on the distribution of carbon isotopes in animals," *Geochim. Cosmochim. Acta* 42, 495–506 (1978).

food intake. That is, the $^{13}C/^{12}C$ in food shows up as an identical $^{13}C/^{12}C$ in the total animal composition, all data would fall on the line in the graph. Typical deviation of the measured values from this behavior, about 0.1%, is very slight and so it seems reasonable to conclude that animal tissue isotopic composition closely resembles that of diet intake. It is interesting to recognize, however, that when there are several layers to a food hierarchy, the fractionation can accumulate. Marine food chains are known that have up to seven tiers implying a heavier carbon isotope fractionation of seven times these values. The 1978 publication has been cited over 3,100 times as of late 2018, attesting to its influence.

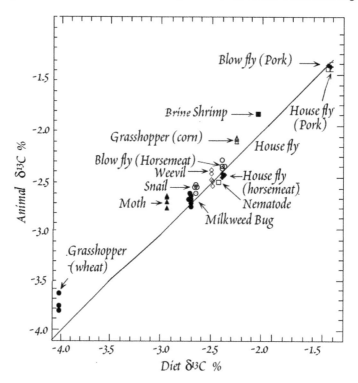

THE VOSTOK SAGA

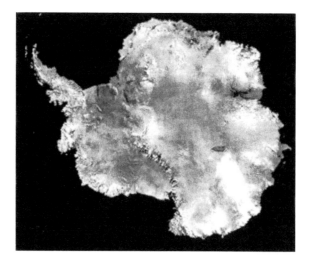

The Antarctic continent shown as a satellite image has surface temperatures averaging about −60°C. Despite that, at a depth of about 4 km below the ice surface, there exists what is currently the world's largest known sub-glacial lake – Lake Vostok, 250 km long and as much as 80 km in width – kept liquid presumably by heat from Earth's interior. The bullet-shaped lake's location is at four o'clock relative to the continent's center. Airborne radar suggests that more than 100 such lakes may be found beneath the Antarctic ice and also that they might be interconnected.

Lake Vostok is apparently several hundred meters deep in its center. The accumulated ice above the lake very slowly melts at its contact with the lake, dropping any extremely small quantities of insoluble matter it might contain to settle to the lake bottom. Since it is estimated that the lake is hundreds of thousands of years old, it is probable that perhaps even hundreds of meters of sediment have deposited over time despite the low deposition rate. The lake itself is the object of a variety of recent experimental probes.

Lake Vostok was discovered accidentally in the 1960s by Russians who had set up a research station* fortuitously located directly above the lake. Earth's record cold was recorded here on July 21, 1983: −89.2°C or −128.6°F. Part of the research involved extracting a 3,000 m ice core (more than one mile) from which carbon dioxide measurements shown next were made on entrapped bubbles covering a period of time estimated to be 160,000 years. The ice core depth was subsequently extended to a 400,000-year equivalent. More recently, another core was cut from the Taylor Dome ice cap near the Ross ice shelf and extended the data, shown below, to 650,000 years. Data from shallower cores are included for times earlier than 40,000 years. Points for carbon dioxide data between about 40,000 years and 400,000 years are from Vostok followed by new, Taylor Dome data on older cores.

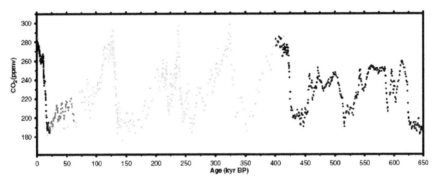

The subsequent figure shows a pair of curves are for the ice hydrogen fractionation ($^2H/^1H$) from two of the drilling sites, Vostok being the bottom tracing. Hydrogen isotopes, with their 100% mass difference, are very sensitive proxies for temperature and, with proper calibration as invented by Urey, can be converted into temperature variations. The hydrogen fractionation trends essentially imply what the temperature trends (not shown here) would be.

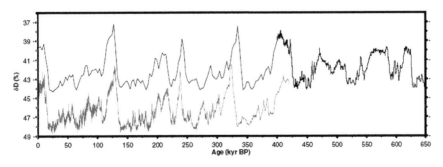

---

* Vostok Research Station was named after the Russian Antarctic explorer ship, the Vostok, from around 1820. The former USSR also named a series of spacecraft after the Vostok. "Vostok" is Russian for "East."

COORDINATED CHRONICLES

Carbonate layers in the ocean bottom sediments can usually be seen very clearly. Here we have a view of a sediment core* from near Bermuda showing alternating color layers, the whitest of which has high $CaCO_3$ content. Darkening gray corresponds to increasing clay content. Rust colored stripes would be due to iron oxide. This particular core was removed from a segment at about 35–45 m depth from the top and corresponds to years covering about 70,000 to 140,000 years ago as determined by radioisotope dating.

Instrumental scans of the Bermuda core were taken to quantify the changes over time on a grayscale. A second core sample was also studied to reinforce the idea that the sediment patterns are representative of the extended geographical region. These scan results are reproduced here for the two cores.

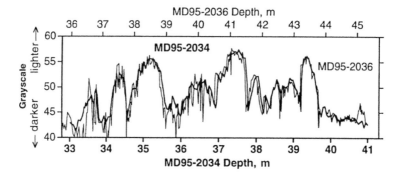

Ultimately, if the Bermuda results are of broader significance, there should be some match with similar measurements made in faraway locations. One such location is associated with a site in central Greenland where ice cores were studied. Compared in the figure are the complete outcomes of oxygen fractionation variations at the Greenland site with the carbonate band (lightness) scans of the full Bermuda core. The various isotopic "stages" are indicated by numbers and make the correlations between the two studies easier to discern. Time in the core may be resolved down to nearly individual centuries over this multi-millennial sampling. Accurate alignment of various time markers is not yet that good though.

---

* E. A. Boyle, "Characteristics of the deep ocean carbon system during the past 150,000 years: ΣCO_2 distributions, deep water flow patterns, and abrupt climate change," *Proc. Nat. Acad. Sci.* 94, 8300 (1997).

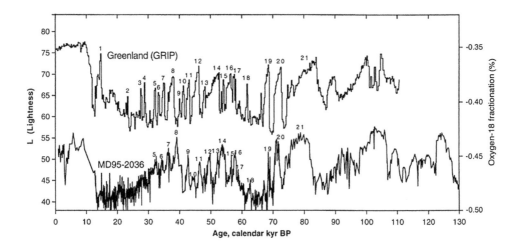

Next, we show the results of measuring $\delta^{18}O$ in such marine carbonates from bottom ocean sediment layers. The graph shows the temperature difference compared to the mean annual temperature (which would read as 0% difference). ΔT was determined from a calibration using Urey's theory of how fractionation mirrors temperature differences employing hydrogen mass ratio proxies in the ice. The peaks and valleys for the buried ocean sediment carbonates correspond extremely well with those found in the Antarctic Vostok ice core measurements for the most recent 160,000 years studied.

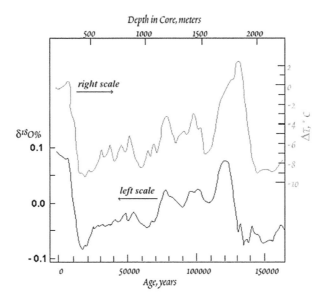

Diametrically across the planet, the Greenland studies give records down to ages of more than 100,000 years, which agree well and correlate in their major features with those found at Vostok. Global changes are implied.

THE DEVIL'S HOLE TALE

Yet another CO_2 legacy story comes from an extensive calcite deposit in Nevada desert country. Only 25 miles from Death Valley lies an area of warm groundwater discharges whose temperature is about 30°C. The most intriguing among these sites is called Devil's Hole, an opening in the ground that is surrounded with calcite formed during 500,000 years* of precipitation from ground-water supersaturated with calcium carbonate. It is also home to the celebrated one inch long *pupfish* whose predicted extinction led to protective legislation by the U.S. Supreme Court in 1976.

Mapped out in the figure, oxygen isotope fractionation in calcium carbonate retrieved by divers from Devil's Hole is compared with ocean bottom sediment carbonate oxygen isotope records ("SPECMAP," dashed lines in the figure). The latter trend has been inverted – flipped over – to make the fluctuation matches more apparent. Age determination in the Devil's Hole measurements was done by careful uranium/thorium/protactinium radioactive chain methods. A very steep drop at about 130,000 years ago (topped by an asterisk in the illustration) corresponds with what is believed to be the termination of the next to the most recent ice age. The results from Devil's Hole presumably represent local conditions, not globally averaged behavior. The marine sediments have isotope fractionations that are influenced mostly by the amount of ^{18}O-depleted water tied up in glacial ice masses. The melting of the ice caps releasing snow-derived ^{16}O-enriched water would drop the ^{18}O content of the ocean. That minimum appears in the inverted display as a maximum (asterisk).

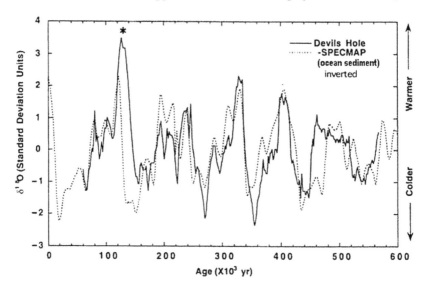

Calling up Vostok's ice oxygen isotopic data[†] shown for comparison to the Devil's Hole next: the synchronized variations are remarkable although some possible displacements of thousands of years could just be a reflection of uncertainties in assigning precise dates.

---

* Age determinations were performed using established uranium and thorium radioactive decay technique from coprecipitated impurities.
† Two Vostok displays correspond to alternative methods for establishing the ages of the ice layers.

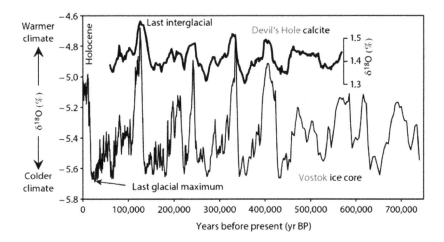

Finally, a minor controversy over possible differences for the penultimate "ice age termination" date at around 130,000 years ago has recently been ameliorated. Comparing three independent results, from Vostok, from ocean sediments, and from the Devil's Hole, initially gave the appearance that fractionations occurred out of synchrony at different locations. But considerably improved radiometric dating instrumentation techniques have allowed measurements of carbonates from Barbados coral and the Devil's Hole carbonates to be revised. They are now essentially concurrent.[*][†] In Chapter 8 on proxies, we saw results of ^{18}O changes in carbonates from a Sunbao cave in China that revealed a stunning shift at 129.0 ± 0.1 thousand years ago congruent with all other markers for the termination.

POSTSCRIPT

The future will yield a rich crop of isotope studies from the Antarctic, from Greenland, from the ocean depths, from terrestrial sites, from mountain chains, and countless other places. Combined with ever improving instrumental techniques and scientists' quest to unravel the conundrums and surprises of nature, much much more will be revealed about the past through studies of carbon dioxide and its avatars.

---

[*] R. L. Edwards, H. Cheng, M. T. Murrell, and S. J. Goldstein, "Protactinium-231 dating of carbonates by thermal ionization mass spectrometry: Implications for quaternary climate change," *Science* 276, 782 (1997).
[†] G. E. Moseley et al., "Reconciliation of the Devils Hole climate record with orbital forcing," *Science* 351, 165 (2016).

Appendix: On a Piece of Chalk

The following appendix reproduces in full the text of Thomas Henry Huxley's "On a piece of chalk: A lecture to working men", published in 1868 in issue 18 of *MacMillan's Magazine*.

Thomas Henry Huxley

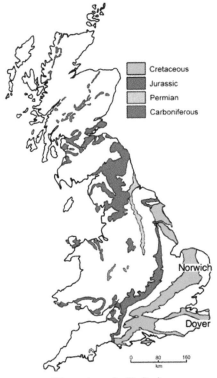

Limestone in Britain

If a well were sunk at our feet in the midst of the city of Norwich, the diggers would very soon find themselves at work in that white substance almost too soft to be called rock, with which we are all familiar as "chalk".

Not only here, but over the whole country of Norfolk, the well-sinker might carry his shaft down many hundred feet without coming to the end of the chalk; and, on the sea-coast, where the waves have pared away the face of the land which breasts them, the scarped faces of the high cliffs are often wholly formed of the same material. Northward, the chalk may be followed as far as Yorkshire; on the south coast it appears abruptly in the picturesque western bays of Dorset, and breaks into the Needles of the Isle of Wight; while on the shores of Kent it supplies that long line of white cliffs to which England owes her name of Albion.

Were the thin soil which covers it all washed away, a curved band of white chalk, here broader, and there narrower, might be followed diagonally across England from Lulworth in Dorset, to Flamborough Head in Yorkshire – a distance of over 280 miles as the crow flies. From this band to the North Sea, on the east, and the Channel, on the south, the chalk is largely hidden by other deposits; but, except in the Weald of Kent and Sussex, it enters into the very foundation of all the southeastern counties.

Attaining, as it does in some places, a thickness of more than a thousand feet, the English chalk must be admitted to be a mass of considerable magnitude. Nevertheless, it covers but an insignificant portion of the whole area occupied by the chalk formation of the globe, much of which has the same general characters as ours, and is found in detached patches, some less, and others more extensive, than the English. Chalk occurs in north-west Ireland; it stretches over a large part of France,–the chalk which underlies Paris being, in fact, a continuation of that of the London basin; it runs through Denmark and Central Europe, and extends southward to North Africa; while eastward, it appears in the Crimea and in Syria, and may be traced as far as the shores of the Sea of Aral, in Central Asia. If all the points at which true chalk occurs were circumscribed, they would lie within an irregular oval about 3,000 miles in long diameter – the area of which would be as great as that of Europe, and would many times exceed that of the largest existing inland sea – the Mediterranean.

Thus the chalk is no unimportant element in the masonry of the earth's crust, and it impresses a peculiar stamp, varying with the conditions to which it is exposed, on the scenery of the districts in which it occurs. The undulating downs and rounded coombs, covered with sweet-grassed turf, of our inland chalk country, have a peacefully domestic and mutton-suggesting prettiness, but can hardly be called either grand or beautiful. But on our southern coasts, the wall-sided cliffs, many hundred feet high, with vast needles and pinnacles standing out in the sea, sharp and solitary enough to serve as perches for the wary cormorant, confer a wonderful beauty and grandeur upon the chalk headlands. And, in the East, chalk has its share in the formation of some of the most venerable of mountain ranges, such as the Lebanon.

What is this wide-spread component of the surface of the earth? and whence did it come?

You may think this no very hopeful inquiry. You may not unnaturally suppose that the attempt to solve such problems as these can lead to no result, save that of entangling the inquirer in vague speculations, incapable of refutation and of verification. If such were really the case, I should have selected some other subject than a "piece of chalk" for my discourse. But, in truth, after much deliberation, I have been unable to think of any topic which would so well enable me to lead you to see how solid is the foundation upon which some of the most startling conclusions of physical science rest.

A great chapter of the history of the world is written in the chalk. Few passages in the history of man can be supported by such an overwhelming mass of direct and indirect evidence as that which testifies to the truth of the fragment of the history of the globe, which I hope to enable you to read, with your own eyes, to-night. Let me add, that few chapters of human history have a more profound significance for ourselves. I weigh my words well when I assert, that the man who should know the true history of the bit of chalk which every carpenter carries about in his breeches-pocket, though ignorant of all other history, is likely, if he will think his knowledge out to its ultimate results, to have a truer, and therefore a better, conception of this wonderful universe, and of man's relation to it, than the most learned student who is deep-read in the records of humanity and ignorant of those of Nature.

The language of the chalk is not hard to learn, not nearly so hard as Latin, if you only want to get at the broad features of the story it has to tell; and I propose that we now set to work to spell that story out together.

We all know that if we "burn" chalk the result is quick-lime. Chalk, in fact, is a compound of carbonic acid gas, and lime, and when you make it very hot the carbonic acid flies away and the lime is left. By this method of procedure we see the lime, but we do not see the carbonic acid. If, on the other hand, you were to powder a little chalk and drop it into a good deal of strong vinegar, there would be a great bubbling and fizzing, and, finally, a clear liquid, in which no sign of chalk would appear. Here you see the carbonic acid in the bubbles; the lime, dissolved in the vinegar, vanishes from sight. There are a great many other ways of showing that chalk is essentially nothing but carbonic acid and quicklime. Chemists enunciate the result of all the experiments which prove this, by stating that chalk is almost wholly composed of "carbonate of lime".

It is desirable for us to start from the knowledge of this fact, though it may not seem to help us very far toward what we seek. For carbonate of lime is a widely-spread substance, and is met with under very various conditions. All sorts of limestones are composed of more or less pure carbonate of lime. The crust which is often deposited by waters which have drained through limestone rocks, in the form of what are called stalagmites and stalactites, is carbonate of lime. Or, to take a more familiar example, the fur on the inside of a tea-kettle is carbonate of lime; and, for anything chemistry tells us to the contrary, the chalk might be a kind of gigantic fur upon the bottom of the earth-kettle, which is kept pretty hot below.

Let us try another method of making the chalk tell us its own history. To the unassisted eye chalk looks simply like a very loose and open kind of stone. But it is possible to grind a slice of chalk down so thin that you can see through it–until it is thin enough, in fact, to be examined with any magnifying power that may be thought desirable. A thin slice of the fur of a kettle might be made in the same way. If it were examined microscopically, it would show itself to be a more or less distinctly laminated mineral substance, and nothing more.

But the slice of chalk presents a totally different appearance when placed under the microscope. The general mass of it is made up of very minute granules; but, imbedded in this matrix, are innumerable bodies, some smaller and some larger, but, on a rough average, not more than a hundredth of an inch in diameter, having a well-defined shape and structure. A cubic inch of some specimens of chalk may contain hundreds of thousands of these bodies, compacted together with incalculable millions of the granules.

The examination of a transparent slice gives a good notion of the manner in which the components of the chalk are arranged, and of their relative proportions. But, by rubbing up some chalk with a brush in water and then pouring off the milky fluid, so as to obtain sediments of different degrees of fineness, the granules and the minute rounded bodies may be pretty well separated from one another, and submitted to microscopic examination, either as opaque or as transparent objects. By combining the views obtained in these various methods, each of the rounded bodies may be proved to be a beautifully-constructed calcareous fabric, made up of a number of chambers, communicating freely with one another. The chambered bodies are of various forms. One of the commonest is something like a badly-grown raspberry, being formed of a number of nearly globular chambers of different sizes congregated together. It is called *Globigerina,* and some specimens of chalk consist of little else than *Globigerinæ* and granules. Let us fix our attention upon the *Globigerina*. It is the spoor of the game we are tracking. If we can learn what it is and what are the conditions of its existence, we shall see our way to the origin and past history of the chalk.

A suggestion which may naturally enough present itself is, that these curious bodies are the result of some process of aggregation which has taken place in the carbonate of lime; that, just as in winter, the rime on our windows simulates the most delicate and elegantly arborescent foliage–proving that the mere mineral water may, under certain conditions, assume the outward form of organic bodies–so this mineral substance, carbonate of lime, hidden away in the bowels of the earth, has taken the shape of these chambered bodies. I am not raising a merely fanciful and unreal objection. Very learned men, in former days, have even entertained the notion that all the formed things found in rocks are of this nature; and if no such conception is at present held to be admissible, it is because long and varied experience has now shown that mineral matter never does assume the form and structure we find in fossils. If any one were to try to persuade you that an oyster-shell (which is also chiefly composed of carbonate of lime) had crystallized out of sea-water I suppose you would laugh at the absurdity. Your laughter would be justified by the fact that all experience tends to show that oyster-shells are formed by the agency of oysters, and in no other way. And if there were no better reasons, we should be justified, on like grounds, in believing that *Globigerina* is not the product of anything but vital activity.

Happily, however, better evidence in proof of the organic nature of the *Globigerinæ* than that of analogy is forthcoming. It so happens that calcareous skeletons, exactly similar to the *Globigerinæ* of the chalk, are being formed, at the present moment, by minute living creatures, which flourish in multitudes, literally more numerous than the sands of the sea-shore, over a large extent of that part of the earth's surface which is covered by the ocean.

The history of the discovery of these living *Globigerinæ,* and of the part which they play in rock building, is singular enough. It is a discovery which, like others of no less scientific importance, has arisen, incidentally, out of work devoted to very different and exceedingly practical interests. When men first took to the sea, they speedily learned to look out for shoals and rocks; and the more the burthen of their ships increased, the more imperatively necessary it became for sailors to ascertain with precision the depth of the waters they traversed. Out of this necessity grew the use of the lead and sounding line; and, ultimately, marine-surveying, which is the recording of the form of coasts and of the depth of the sea, as ascertained by the sounding lead, upon charts.

At the same time, it became desirable to ascertain and to indicate the nature of the sea-bottom, since this circumstance greatly affects its goodness as holding ground for anchors. Some ingenious tar, whose name deserves a better fate than the oblivion into which it has fallen, attained this object by "arming" the bottom of the lead with a lump of grease, to which more or less of the sand or mud, or broken shells, as the case might be, adhered, and was brought to the surface. But, however well adapted such an apparatus might be for rough nautical purposes, scientific accuracy could not be expected from the armed lead, and to remedy its defects (especially when applied to sounding in great depths) Lieut. Brooke, of the American Navy, some years ago invented a most ingenious machine, by which a considerable portion of the superficial layer of the sea-bottom can be scooped out and brought up from any depth to which the lead descends. In 1853, Lieut. Brooke obtained mud from the bottom of the North Atlantic, between Newfoundland and the Azores, at a depth of more than 10,000 feet, or two miles, by the help of this sounding apparatus. The specimens were sent for examination to Ehrenberg of Berlin, and to Bailey of West Point, and those able microscopists found that this deep-sea mud was almost entirely composed of the skeletons of living organisms–the greater proportion of these being just like the *Globigerinæ* already known to occur in the chalk.

Thus far, the work had been carried on simply in the interest of science, but Lieut. Brooke's method of sounding acquired a high commercial value, when the enterprise of laying down the telegraph-cable between this country and the United States was undertaken. For it became a matter of immense importance to know, not only the depth of the sea over the whole line along which the cable was to be laid, but the exact nature of the bottom, so as to guard against chances of cutting or fraying the strands of that costly rope. The Admiralty consequently ordered Captain Dayman, an old friend and shipmate of mine, to ascertain the depth over the whole line of the cable, and to bring back specimens of the bottom. In former days, such a command as this might have sounded very much like one of the impossible things which the young Prince in the Fairy Tales is ordered to do before he can obtain the hand of the Princess. However, in the months of June and July, 1857, my friend performed the task assigned to him with great expedition and precision, without, so far as I know, having met with any reward of that kind. The specimens of Atlantic mud which he procured were sent to me to be examined and reported upon.*

The result of all these operations is, that we know the contours and the nature of the surface-soil covered by the North Atlantic for a distance of 1,700 miles from east to west, as well as we know that of any part of the dry land. It is a prodigious plain – one of the widest and most even plains in the world. If the sea were drained off, you might drive a waggon all the way from Valentia, on the west coast of Ireland, to Trinity Bay, in Newfoundland. And, except upon one sharp incline about 200 miles from Valentia, I am not quite sure that it would even be necessary to put the skid on, so gentle are the ascents and descents upon that long route. From Valentia the road would lie down-hill for about 200 miles to the point at which the bottom is now covered by 1,700 fathoms of sea-water. Then would come the central plain, more than a thousand miles wide, the inequalities of the surface of which would be hardly perceptible, though the depth of water upon it now varies from 10,000 to 15,000 feet; and there are places in which Mont Blanc might be sunk without showing its peak above water. Beyond this, the ascent on the American side commences, and gradually leads, for about 300 miles, to the Newfoundland shore.

Almost the whole of the bottom of this central plain (which extends for many hundred miles in a north and south direction) is covered by a fine mud, which, when brought to the surface, dries into a greyish white friable substance. You can write with this on a blackboard, if you are so inclined; and, to the eye, it is quite like very soft, greyish chalk. Examined chemically, it proves to be composed almost wholly of carbonate of lime; and if you make a section of it, in the same way as that of the piece of chalk was made, and view it with the microscope, it presents innumerable *Globigerinæ* imbedded in a granular matrix. Thus this deep-sea mud is substantially chalk. I say substantially, because there are a good many minor differences; but as these have no bearing on the question immediately before us, – which is the nature of the *Globigerinæ* of the chalk, – it is unnecessary to speak of them.

Globigerinæ of every size, from the smallest to the largest, are associated together in the Atlantic mud, and the chambers of many are filled by a soft animal matter. This soft substance is, in fact, the remains of the creature to which the *Globigerina* shell, or rather skeleton, owes its existence and which is an animal of the simplest imaginable description. It is, in fact, a

---

* See Appendix to Captain Dayman's, *Deep-Sea Soundings in the North Atlantic Ocean between Ireland and Newfoundland, Made in H.M.S. "Cyclops."* Published by order of the Lords Commissioners of the Admiralty, 1858. They have since formed the subject of an elaborate Memoir by Messrs. Parker and Jones, published in the *Philosophical Transactions* for 1865.

mere particle of living jelly, without defined parts of any kind – without a mouth, nerves, muscles, or distinct organs, and only manifesting its vitality to ordinary observation by thrusting out and retracting from all parts of its surface, long filamentous processes, which serve for arms and legs. Yet this amorphous particle, devoid of everything which, in the higher animals, we call organs, is capable of feeding, growing, and multiplying; of separating from the ocean the small proportion of carbonate of lime which is dissolved in sea-water; and of building up that substance into a skeleton for itself, according to a pattern which can be imitated by no other known agency.

The notion that animals can live and flourish in the sea, at the vast depths from which apparently living *Globigerinæ* have been brought up, does not agree very well with our usual conceptions respecting the conditions of animal life; and it is not so absolutely impossible as it might at first sight appear to be, that the *Globigerinæ* of the Atlantic sea-bottom do not live and die where they are found.

As I have mentioned, the soundings from the great Atlantic plain are almost entirely made up of *Globigerinæ* with the granules which have been mentioned, and some few other calcareous shells; but a small percentage of the chalky mud – perhaps at most some 5% of it – is of a different nature, and consists of shells and skeletons composed of silex, or pure flint. These silicious bodies belong partly to the lowly vegetable organisms which are called *Diatomaceæ*, and partly to the minute, and extremely simple, animals, termed *Radiolaria*. It is quite certain that these creatures do not live at the bottom of the ocean, but at its surface – where they may be obtained in prodigious numbers by the use of a properly constructed net. Hence it follows that these silicious organisms, though they are not heavier than the lightest dust, must have fallen, in some cases, through fifteen thousand feet of water, before they reached their final resting-place on the ocean floor. And considering how large a surface these bodies expose in proportion to their weight, it is probable that they occupy a great length of time in making their burial journey from the surface of the Atlantic to the bottom.

But if the *Radiolaria* and Diatoms are thus rained upon the bottom of the sea, from the superficial layer of its waters in which they pass their lives, it is obviously possible that the *Globigerinæ* may be similarly derived; and if they were so, it would be much more easy to understand how they obtain their supply of food than it is at present. Nevertheless, the positive and negative evidence all points the other way. The skeletons of the full-grown, deep sea *Globigerinæ* are so remarkably solid and heavy in proportion to their surface as to seem little fitted for floating; and, as a matter of fact, they are not to be found along with the Diatoms and *Radiolaria* in the uppermost stratum of the open ocean. It has been observed, again, that the abundance of *Globigerinæ,* in proportion to other organisms, of like kind, increases with the depth of the sea; and that deep water *Globigerinæ* are larger than those which live in shallower parts of the sea; and such facts negative the supposition that these organisms have been swept by currents from the shallows into the deeps of the Atlantic. It therefore seems to be hardly doubtful that these wonderful creatures live and die at the depths in which they are found.*

However, the important points for us are, that the living *Globigerinæ* are exclusively marine animals, the skeletons of which abound at the bottom of deep seas; and that there is not a shadow of reason for believing that the habits of the *Globigerinæ* of the chalk differed

---

* During the cruise of H.M.S. *Bulldog,* commanded by Sir Leopold McClintock, in 1860, living star-fish were brought up, clinging to the lowest part of the sounding-line, from a depth of 1,260 fathoms, midway between Cape Farewell, in Greenland, and the Rockall banks. Dr. Wallich ascertained that the sea-bottom at this point consisted of the ordinary *Globigerina* ooze, and that the stomachs of the star-fishes were full of *Globigerinæ*. This discovery removes all objections to the existence of living *Globigerinæ* at great depths, which are based upon the supposed difficulty of maintaining animal life under such conditions; and it throws the burden of proof upon those who object to the supposition that the *Globigerinæ* live and die where they are found.

from those of the existing species. But if this be true, there is no escaping the conclusion that the chalk itself is the dried mud of an ancient deep sea.

In working over the soundings collected by Captain Dayman, I was surprised to find that many of what I have called the "granules" of that mud were not, as one might have been tempted to think at first, the mere powder and waste of *Globigerinæ*, but that they had a definite form and size. I termed these bodies "*coccoliths*", and doubted their organic nature. Dr. Wallich verified my observation, and added the interesting discovery that, not unfrequently, bodies similar to these "coccoliths" were aggregated together into spheroids, which he termed "*coccospheres*". So far as we knew, these bodies, the nature of which is extremely puzzling and problematical, were peculiar to the Atlantic soundings. But, a few years ago, Mr. Sorby, in making a careful examination of the chalk by means of thin sections and otherwise, observed, as Ehrenberg had done before him, that much of its granular basis possesses a definite form. Comparing these formed particles with those in the Atlantic soundings, he found the two to be identical; and thus proved that the chalk, like the surroundings, contains these mysterious coccoliths and coccospheres. Here was a further and most interesting confirmation, from internal evidence, of the essential identity of the chalk with modern deep-sea mud. *Globigerinæ*, coccoliths, and coccospheres are found as the chief constituents of both, and testify to the general similarity of the conditions under which both have been formed.*

The evidence furnished by the hewing, facing, and superposition of the stones of the Pyramids, that these structures were built by men, has no greater weight than the evidence that the chalk was built by *Globigerinæ* and the belief that those ancient pyramid-builders were terrestrial and air-breathing creatures like ourselves, is not better based than the conviction that the chalk-makers lived in the sea. But as our belief in the building of the Pyramids by men is not only grounded on the internal evidence afforded by these structures, but gathers strength from multitudinous collateral proofs and is clinched by the total absence of any reason for a contrary belief; so the evidence drawn from the *Globigerinæ* that the chalk is an ancient sea-bottom; is fortified by innumerable independent lines of evidence; and our belief in the truth of the conclusion to which all positive testimony tends, receives the like negative justification from the fact that no other hypothesis has a shadow of foundation.

It may be worth while briefly to consider a few of these collateral proofs that the chalk was deposited at the bottom of the sea. The great mass of the chalk is composed, as we have seen, of the skeletons of *Globigerinæ*, and other simple organisms, imbedded in granular matter. Here and there, however, this hardened mud of the ancient sea reveals the remains of higher animals which have lived and died, and left their hard parts in the mud, just as the oysters die and leave their shells behind them, in the mud of the present seas.

There are, at the present day, certain groups of animals which are never found in fresh waters, being unable to live anywhere but in the sea. Such are the corals; those corallines which are called *Polyzoa;* those creatures which fabricate the lamp-shells, and are called *Brachiopoda;* the pearly *Nautilus* and all animals allied to it; and all the forms of sea-urchins and star-fishes. Not only are all these creatures confined to salt water at the present day; but, so far as our records of the past go, the conditions of their existence have been the same: hence, their occurrence in any deposit is as strong evidence as can be obtained, that that deposit was formed in the sea. Now the remains of animals of all kinds which have been enumerated, occur in the chalk, in greater or less abundance; while not one of those forms of shell-fish which are characteristic of fresh water has yet been observed in it.

---

* I have recently traced out the development of the "coccoliths" from a diameter of 1/7000th of an inch up to their largest size (which is about 1/6000th), and no longer doubt that they are produced by independent organisms, which, like the *Globigerinæ*, live and die at the bottom of the sea.

When we consider that the remains of more than three thousand distinct species of aquatic animals have been discovered among the fossils of the chalk, that the great majority of them are of such forms as are now met with only in the sea, and that there is no reason to believe that any one of them inhabited fresh water–the collateral evidence that the chalk represents an ancient sea-bottom acquires as great force as the proof derived from the nature of the chalk itself. I think you will now allow that I did not overstate my case when I asserted that we have as strong grounds for believing that all the vast area of dry land, at present occupied by the chalk, was once at the bottom of the sea, as we have for any matter of history whatever; while there is no justification for any other belief.

No less certain it is that the time during which the countries we now call south-east England, France, Germany, Poland, Russia, Egypt, Arabia, Syria, were more or less completely covered by a deep sea, was of considerable duration. We have already seen that the chalk is, in places, more than a thousand feet thick. I think you will agree with me, that it must have taken some time for the skeletons of animalcules of a hundredth of an inch in diameter to heap up such a mass as that. I have said that throughout the thickness of the chalk the remains of other animals are scattered. These remains are often in the most exquisite state of preservation. The valves of the shell-fishes are commonly adherent; the long spines of some of the sea-urchins, which would be detached by the smallest jar, often remain in their places. In a word, it is certain that these animals have lived and died when the place which they now occupy was the surface of as much of the chalk as had then been deposited; and that each has been covered up by the layer of *Globigerina* mud, upon which the creatures imbedded a little higher up have, in like manner, lived and died. But some of these remains prove the existence of reptiles of vast size in the chalk sea. These lived their time, and had their ancestors and descendants, which assuredly implies time, reptiles being of slow growth.

There is more curious evidence, again, that the process of covering up, or, in other words, the deposit of *Globigerina* skeletons, did not go on very fast. It is demonstrable that an animal of the cretaceous sea might die, that its skeleton might lie uncovered upon the sea-bottom long enough to lose all its outward coverings and appendages by putrefaction; and that, after this had happened, another animal might attach itself to the dead and naked skeleton, might grow to maturity, and might itself die before the calcareous mud had buried the whole.

Cases of this kind are admirably described by Sir Charles Lyell. He speaks of the frequency with which geologists find in the chalk a fossilized sea urchin, to which is attached the lower valve of a *Crania*. This is a kind of shell-fish, with a shell composed of two pieces, of which, as in the oyster, one is fixed and the other free.

"The upper valve is almost invariably wanting, though occasionally found in a perfect state of preservation in the white chalk at some distance. In this case, we see clearly that the sea-urchin first lived from youth to age, then died and lost its spines, which were carried away. Then the young *Crania* adhered to the bared shell, grew and perished in its turn; after which, the upper valve was separated from the lower, before the Echinus became enveloped in chalky mud".*

A specimen in the Museum of Practical Geology, in London, still further prolongs the period which must have elapsed between the death of the sea-urchin, and its burial by the *Globigerinæ*. For the outward face of the valve of a *Crania,* which is attached to a sea urchin (*Micraster*), is itself overrun by an incrusting coralline, which spreads thence over more or less of the surface of the sea urchin. It follows that, after the upper valve of the *Crania* fell off, the surface of the attached valve must have remained exposed long enough to allow of the growth of the whole coralline, since corallines do not live embedded in mud.

---

* *Elements of Geology,* by Sir Charles Lyell, Bart., F.R.S., p. 23.

The progress of knowledge may, one day, enable us to deduce from such facts as these the maximum rate at which the chalk can have accumulated and thus to arrive at the minimum duration of the chalk period. Suppose that the valve of the *Crania* upon which a coralline has fixed itself in the way just described is so attached to the sea urchin that no part of it is more than an inch above the face upon which the sea urchin rests. Then, as the coralline could not have fixed itself if the *Crania* had been covered up with chalk mud and could not have lived had itself been so covered, it follows that an inch of chalk mud could not have accumulated within the time between the death and decay of the soft parts of the sea urchin and the growth of the coralline to the full size which it has attained. If the decay of the soft parts of the sea-urchin; the attachment, growth to maturity, and decay of the *Crania*; and the subsequent attachment and growth of the coralline, took a year (which is a low estimate enough), the accumulation of the inch of chalk must have taken more than a year; and the deposit of a thousand feet of chalk must, consequently, have taken more than twelve thousand years.

The foundation of all this calculation is, of course, a knowledge of the length of time the *Crania* and the coralline needed to attain their full size, and, on this head, precise knowledge is at present wanting. But there are circumstances which tend to show that nothing like an inch of chalk has accumulated during the life of a *Crania*, and, on any probable estimate of the length of that life, the chalk period must have had a much longer duration than that thus roughly assigned to it.

Thus, not only is it certain that the chalk is the mud of an ancient sea-bottom; but it is no less certain, that the chalk sea existed during an extremely long period, though we may not be prepared to give a precise estimate of the length of that period in years. The relative duration is clear, though the absolute duration may not be definable. The attempt to affix any precise date to the period at which the chalk sea began, or ended, its existence, is baffled by difficulties of the same kind. But the relative age of the cretaceous epoch may be determined with as great ease and certainty as the long duration of that epoch.

You will have heard of the interesting discoveries recently made, in various parts of Western Europe, of flint implements, obviously worked into shape by human hands, under circumstances which show conclusively that man is a very ancient denizen of these regions. It has been proved that the whole populations of Europe, whose existence has been revealed to us in this way, consisted of savages, such as the Esquimaux are now; that, in the country which is now France, they hunted the reindeer, and were familiar with the ways of the mammoth and the bison. The physical geography of France was in those days different from what it is now – the river Somme, for instance, having cut its bed a hundred feet deeper between that time and this; and, it is probable, that the climate was more like that of Canada or Siberia, than that of Western Europe.

The existence of these people is forgotten even in the traditions of the oldest historical nations. The name and fame of them had utterly vanished until a few years back; and the amount of physical change which has been effected since their day renders it more than probable that, venerable as are some of the historical nations, the workers of the chipped flints of Hoxne or of Amiens are to them, as they are to us, in point of antiquity. But, if we assign to these hoar relics of long-vanished generations of men the greatest age that can possibly be claimed for them, they are not older than the drift, or boulder clay, which, in comparison with the chalk, is but a very juvenile deposit. You need go no further than your own sea-board for evidence of this fact. At one of the most charming spots on the coast of Norfolk, Cromer, you will see the boulder clay forming a vast mass, which lies upon the chalk, and must consequently have come into existence after it. Huge boulders of chalk are, in fact included in the clay, and have evidently been brought to the position they now occupy by the same agency as that which has planted blocks of syenite from Norway side by side with them.

The chalk, then, is certainly older than the boulder clay. If you ask how much, I will again take you no further than the same spot upon your own coasts for evidence. I have spoken of the boulder clay and drift as resting upon the chalk. That is not strictly true. Interposed between the chalk and the drift is a comparatively insignificant layer, containing vegetable matter. But that layer tells a wonderful history. It is full of stumps of trees standing as they grew. Fir-trees are there with their cones, and hazel-bushes with their nuts; there stand the stools of oak and yew trees, beeches and alders. Hence this stratum is appropriately called the "forest-bed".

It is obvious that the chalk must have been upheaved and converted into dry land, before the timber trees could grow upon it. As the bolls of some of these trees are from two to three feet in diameter, it is no less clear that the dry land thus formed remained in the same condition for long ages. And not only do the remains of stately oaks and well-grown firs testify to the duration of this condition of things, but additional evidence to the same effect is afforded by the abundant remains of elephants, rhinoceroses, hippopotamuses, and other great wild beasts, which it has yielded to the zealous search of such men as the Rev. Mr. Gunn. When you look at such a collection as he has formed, and bethink you that these elephantine bones did veritably carry their owners about, and these great grinders crunch, in the dark woods of which the forest-bed is now the only trace, it is impossible not to feel that they are as good evidence of the lapse of time as the annual rings of the tree stumps.

Thus there is a writing upon the wall of cliffs at Cromer, and whoso runs may read it. It tells us, with an authority which cannot be impeached, that the ancient sea bed of the chalk sea was raised up, and remained dry land, until it was covered with forest, stocked with the great game the spoils of which have rejoiced your geologists. How long it remained in that condition cannot be said; but, "the whirligig of time brought its revenges" in those days as in these. That dry land, with the bones and teeth of generations of long-lived elephants, hidden away among the gnarled roots and dry leaves of its ancient trees, sank gradually to the bottom of the icy sea, which covered it with huge masses of drift and boulder clay. Sea-beasts, such as the walrus now restricted to the extreme north, paddled about where birds had twittered among the topmost twigs of the fir-trees. How long this state of things endured we know not, but at length it came to an end. The upheaved glacial mud hardened into the soil of modern Norfolk. Forests grew once more, the wolf and the beaver replaced the reindeer and the elephant; and at length what we call the history of England dawned.

Thus you have, within the limits of your own county, proof that the chalk can justly claim a very much greater antiquity than even the oldest physical traces of mankind. But we may go further and demonstrate, by evidence of the same authority as that which testifies to the existence of the father of men, that the chalk is vastly older than Adam himself. The Book of Genesis informs us that Adam, immediately upon his creation, and before the appearance of Eve, was placed in the Garden of Eden. The problem of the geographical position of Eden has greatly vexed the spirits of the learned in such matters, but there is one point respecting which, so far as I know, no commentator has ever raised a doubt. This is, that of the four rivers which are said to run out of it, Euphrates and Hiddekel are identical with the rivers now known by the names of Euphrates and Tigris. But the whole country in which these mighty rivers take their origin, and through which they run, is composed of rocks which are either of the same age as the chalk, or of later date. So that the chalk must not only have been formed, but, after its formation, the time required for the deposit of these later rocks, and for their upheaval into dry land, must have elapsed, before the smallest brook which feeds the swift stream of "the great river, the river of Babylon" began to flow.

Thus, evidence which cannot be rebutted, and which need not be strengthened, though if time permitted I might indefinitely increase its quantity, compels you to believe that the earth, from the time of the chalk to the present day, has been the theatre of a series of changes as vast in their amount, as they were slow in their progress. The area on which we stand has been first sea and then land, for at least four alternations; and has remained in each of these conditions for a period of great length.

Nor have these wonderful metamorphoses of sea into land, and of land into sea, been confined to one corner of England. During the chalk period, or "cretaceous epoch", not one of the present great physical features of the globe was in existence. Our great mountain ranges, Pyrenees, Alps, Himalayas, Andes, have all been upheaved since the chalk was deposited, and the cretaceous sea flowed over the sites of Sinai and Ararat. All this is certain, because rocks of cretaceous, or still later, date have shared in the elevatory movements which gave rise to these mountain chains; and may be found perched up, in some cases, many thousand feet high upon their flanks. And evidence of equal cogency demonstrates that, though, in Norfolk, the forest-bed rests directly upon the chalk, yet it does so, not because the period at which the forest grew immediately followed that at which the chalk was formed, but because an immense lapse of time, represented elsewhere by thousands of feet of rock, is not indicated at Cromer.

I must ask you to believe that there is no less conclusive proof that a still more prolonged succession of similar changes occurred, before the chalk was deposited. Nor have we any reason to think that the first term in the series of these changes is known. The oldest sea-beds preserved to us are sands, and mud, and pebbles, the wear and tear of rocks which were formed in still older oceans.

But, great as is the magnitude of these physical changes of the world, they have been accompanied by a no less striking series of modifications in its living inhabitants. All the great classes of animals, beasts of the field, fowls of the air, creeping things, and things which dwell in the waters, flourished upon the globe long ages before the chalk was deposited. Very few, however, if any, of these ancient forms of animal life were identical with those which now live. Certainly not one of the higher animals was of the same species as any of those now in existence. The beasts of the field, in the days before the chalk, were not our beasts of the field, nor the fowls of the air such as those which the eye of man has seen flying, unless his antiquity dates infinitely further back than we at present surmise. If we could be carried back into those times, we should be as one suddenly set down in Australia before it was colonized. We should see mammals, birds, reptiles, fishes, insects, snails, and the like, clearly recognizable as such, and yet not one of them would be just the same as those with which we are familiar, and many would be extremely different.

From that time to the present, the population of the world has undergone slow and gradual, but incessant, changes. There has been no grand catastrophe–no destroyer has swept away the forms of life of one period, and replaced them by a totally new creation: but one species has vanished and another has taken its place; creatures of one type of structure have diminished, those of another have increased, as time has passed on. And thus, while the differences between the living creatures of the time before the chalk and those of the present day appear startling, if placed side by side, we are led from one to the other by the most gradual progress, if we follow the course of Nature through the whole series of those relics of her operations which she has left behind. It is by the population of the chalk sea that the ancient and the modern inhabitants of the world are most completely connected. The groups which are dying out flourish, side by side, with the groups which are now the dominant forms of life. Thus the chalk contains remains of those strange flying and swimming reptiles, the pterodactyl, the ichthyosaurus and the plesiosaurus, which are found in no later deposits, but abounded in preceding ages. The chambered shells called ammonites and belemnites, which are so characteristic of the period preceding the cretaceous, in like manner die with it.

But, among these fading remainders of a previous state of things, are some very modern forms of life, looking like Yankee pedlars among a tribe of Red Indians. Crocodiles of modern type appear; bony fishes, many of them very similar to existing species, almost supplant the forms of fish which predominate in more ancient seas; and many kinds of living shellfish first become known to us in the chalk. The vegetation acquires a modern aspect. A few living animals are not even distinguishable as species, from those which existed at that remote epoch. The *Globigerina* of the present day, for example, is not different specifically from that of the chalk; and the same may be said of many other *Foraminifera*. I think it probable that critical and unprejudiced examination will show that more than one species of much higher animals have had a similar longevity; but the only example which I can at present give confidently is the snake's-head lamp-shell *(Terebratulina caput serpentis)*, which lives in our English seas and abounded (as *Terebratulina striata* of authors) in the chalk.

The longest line of human ancestry must hide its diminished head before the pedigree of this insignificant shellfish. We Englishmen are proud to have an ancestor who was present at the Battle of Hastings. The ancestors of *Terebratulina caput serpentis* may have been present at a battle of *Ichthyosauria* in that part of the sea which, when the chalk was forming, flowed over the site of Hastings. When all around has changed, this *Terebratulina* has peacefully propagated its species from generation to generation, and stands to this day, as a living testimony to the continuity of the present with the past history of the globe.

Up to this moment I have stated, so far as I know, nothing but well-authenticated facts, and the immediate conclusions which they force upon the mind. But the mind is so constituted that it does not willingly rest in facts and immediate causes, but seeks always after a knowledge of the remoter links in the chain of causation.

Taking the many changes of any given spot of the earth's surface, from sea to land and from land to sea, as an established fact, we cannot refrain from asking ourselves how these changes have occurred. And when we have explained them – as they must be explained – by the alternate slow movements of elevation and depression which have affected the crust of the earth, we go still further back, and ask, Why these movements?

I am not certain that any one can give you a satisfactory answer to that question. Assuredly I cannot. All that can be said, for certain, is, that such movements are part of the ordinary course of nature, inasmuch as they are going on at the present time. Direct proof may be given, that some parts of the land of the northern hemisphere are at this moment insensibly rising and others insensibly sinking; and there is indirect, but perfectly satisfactory, proof, that an enormous area now covered by the Pacific has been deepened thousands of feet, since the present inhabitants of that sea came into existence. Thus there is not a shadow of a reason for believing that the physical changes of the globe, in past times, have been affected by other than natural causes. Is there any more reason for believing that the concomitant modifications in the forms of the living inhabitants of the globe have been brought about in other ways?

Before attempting to answer this question, let us try to form a distinct mental picture of what has happened in some special case. The crocodiles are animals which, as a group, have a very vast antiquity. They abounded ages before the chalk was deposited; they throng the rivers in warm climates, at the present day. There is a difference in the form of the joints of the backbone, and in some minor particulars, between the crocodiles of the present epoch and those which lived before the chalk; but, in the cretaceous epoch, as I have already mentioned, the crocodiles had assumed the modern type of structure. Notwithstanding this, the crocodiles of the chalk are not identically the same as those which lived in the times called "older tertiary", which succeeded the cretaceous epoch; and the crocodiles of the older tertiaries are not identical with those of the newer tertiaries, nor are these identical with existing forms. I leave open the question whether particular species may have lived on from epoch to epoch. But each epoch has had its peculiar crocodiles; though all, since the chalk, have belonged to the modern type, and differ simply in their proportions, and in such structural particulars as are discernible only to trained eyes.

How is the existence of this long succession of different species of crocodiles to be accounted for? Only two suppositions seem to be open to us—Either each species of crocodile has been specially created, or it has arisen out of some pre-existing form by the operation of natural causes. Choose your hypothesis; I have chosen mine. I can find no warranty for believing in the distinct creation of a score of successive species of crocodiles in the course of countless ages of time. Science gives no countenance to such a wild fancy; nor can even the perverse ingenuity of a commentator pretend to discover this sense, in the simple words in which the writer of Genesis records the proceedings of the fifth and sixth days of the Creation.

On the other hand, I see no good reason for doubting the necessary alternative, that all these varied species have been evolved from pre-existing crocodilian forms, by the operation of causes as completely a part of the common order of nature as those which have effected the changes of the inorganic world. Few will venture to affirm that the reasoning which applies to crocodiles loses its force among other animals, or among plants. If one series of species has come into existence by the operation of natural causes, it seems folly to deny that all may have arisen in the same way.

A small beginning has led us to a great ending. If I were to put the bit of chalk with which we started into the hot but obscure flame of burning hydrogen, it would presently shine like the sun. It seems to me that this physical metamorphosis is no false image of what has been the result of our subjecting it to a jet of fervent, though nowise brilliant, thought to-night. It has become luminous, and its clear rays, penetrating the abyss of the remote past, have brought within our ken some stages of the evolution of the earth. And in the shifting "without haste, but without rest" of the land and sea, as in the endless variation of the forms assumed by living beings, we have observed nothing but the natural product of the forces originally possessed by the substance of the universe.

Bibliography

R. Alley, *The Two-Mile Time Machine*, Princeton University Press, Princeton, NJ (1999).

D. Archer and P. Martin, "Thin walls tell the tale," *Science*, 294, 2108 (2001).

S. Arrhenius, "On the influence of carbonic acid in the air upon the temperature of the ground," *Phil. Mag. S.5*, 41, 237–276 (1896).

S. Arrhenius, *Worlds in the Making*, Harper & Bros., New York (1908).

M. A. Arthur, "Volcanic contributions to the carbon and sulfur geochemical cycles and global change," Sigurdsson (*q.v.*), p. 1045.

D. Beerling, *The Emerald Planet: How Plants Changed Earth's History*, Oxford University Press, Oxford (2007).

D. J. Beerling and F. I. Woodward, "Stomatal density responses to global environmental change," *Adv. Bioclimatol.*, 4, 171 (1996).

M. L. Bender, *Paleoclimate*, Princeton University Press, Princeton, NJ (2013).

R. A. Berner, "The long-term carbon cycle, fossil fuels and atmospheric composition," *Nature*, 426, 323 (2003).

R. A. Berner, D. J. Beerling, R. Dudley, J. M. Robinson, and R. A. Wildman, Jr., "Phanerozoic atmospheric oxygen," *Ann. Rev. Earth Planet. Sci.*, 31, 105–134 (2003).

E. K. Berner, R. A. Berner, and K. L. Moulton, "Plants and mineral weathering: Present and past," *Treat. Geochem.*, 5, 169–188 (2003).

R. E. Blankenship, *Molecular Mechanisms of Photosynthesis*, Blackwell Sciences, Oxford (2002).

W. S. Broecker and E. Clark, "Glacial-to-holocene redistribution of carbonate ion in the deep sea," *Science*, 294, 2152 (2001).

R. G. Bruant, Jr., M. A. Celia, C. A. Peters, and A. J. Guswa, "Long-term underground storage of CO_2," *Env. Sci. Tech.*, 36, 240A–245A (2002).

J. N. Butler, *Ionic Equilibrium*, John Wiley & Sons, New York (1998).

D. E. Canfield, "The early history of atmospheric oxygen," *Annu. Rev. Earth Planet. Sci.*, 33, 1 (2005).

T. M. Cronin, *Paleoclimates*, Columbia University Press, New York (2010).

C. L. Van Dover, *The Ecology of Deep-Sea Hydrothermal Vents*, Princeton University Press, Princeton, NJ (2000).

R. Fortey, *Earth: An Intimate History*, Alfred A. Knopf, New York (2004).

W. G. Glasser, R. A. Northey, and T. P. Schultz, eds., *Lignin: Historical, Biological, and Materials Perspectives*, ACS Symposium Series 742, American Chemical Society, Washington, DC (2000).

A. C. Guyton and J. E. Hall, *Human Physiology and Mechanisms of Disease, 6th ed.*, Saunders, Philadelphia, PA (1997).

I. M. Head, D. M. Jones, and S. R. Larter, "Biological activity in the deep subsurface and the origin of heavy oil," *Nature*, 426, 344 (2003).

H. D. Holland, *The Chemistry of the Atmosphere and Oceans*, Wiley-Interscience, New York (1978).

D. M. Hunten, "Atmospheric evolution of the terrestrial planets," *Science*, 259, 915 (1993).

J. Imbrie and K. P. Imbrie, *Ice Ages: Solving the Mystery*, Enslow Publishers, Short Hills, NJ (1979).

Intergovernmental Panel on Climate Change, *Climate Change 2001: The Scientific Basis*, IPCC, Geneva (2001).

Intergovernmental Panel on Climate Change, *Carbon Dioxide Capture and Storage*, IPCC, Cambridge University Press, Geneva, Cambridge (2005).

H. W. Jannasch, *Geophysical Monograph 91*, Am. Geophys. Union, Washington, DC (1995), pp. 273–296.

J. L. Kirschrink et al., "Paleoproterozoic snowball earth: Extreme climatic and geochemical change and its biological consequences," *Proc. Nat'l. Acad. Sci.*, 97, 1400–1405 (2000).

K. B. Krauskopf and D. K. Bird, *Introduction to Geochemistry, 3rd ed.*, McGraw-Hill, Inc., New York (1995).

D. A. Kring and D. D. Durda, "The day the world burned," *Sci. Am.*, 298, 98 (2003).

K. S. Lackner, "Carbonate chemistry for sequestering fossil carbon," *Ann. Rev. Energy Env.*, 27, 193–232 (2002).

N. Lane, *Oxygen: The Molecule that Made the World*, Oxford University Press, Oxford (2002).

J. S. Levine, T. Bobbe, N. Ray, A. Singh, and R. G. Witt, *Wildland Fires and the Environment: A Global Synthesis*, UNEP/DEIAEW/TR.99-1, Division of Environmental Information, Assessment and Early Warning, UN Environment Programme, Nairobi, Kenya (1999).

N. G. Lewis and S. Sarkanen, eds., *Lignin and Lignan Biosynthesis*, ACS Symposium Series 697, American Chemical Society, Washington, DC (1998).

P. A. Mayewski and F. White, *The Ice Chronicles: The Quest to Understand Global Climate Change*, University Press of New England, Lebanon, NH (2002).

T. Nield, *Supercontinent: Ten Billion Years in the Life of Our Planet*, Harvard University Press, Cambridge, MA (2007).

S. Page et al., "The amount of carbon released from peat and forest fires in Indonesia during 1997," *Nature*, 420, 61–65 (2002).

D. Rapp, *Assessing Climate Change: Temperatures, Solar Radiation, and Heat Balance*, Springer, Berlin Heidelberg (2008).

D. H. Rothman, "Atmospheric carbon dioxide levels for the last 500 million years," *PNAS*, 99, 4167–4171 (2002).

D. L. Royer, "CO_2-forced climate thresholds during the Phanerozoic," *Geochim. Et Cosmochim. Acta.*, 70, 5665–5675 (2006).

J. L. Sarmiento and N. Gruber, "Sinks for anthropogenic carbon," *Phys. Today*, 55, 30 (2002).

J. S. Seewald, "Organic-inorganic interactions in petroleum-producing sedimentary basins," *Nature*, 426, 327 (2003).

H. Sigurdsson, ed., *Encyclopedia of Volcanoes*, Academic Press, San Diego, CA (2000).

V. Smil, *The Earth's Biosphere: Evolution, Dynamics and Change*, MIT Press, Cambridge, MA (2002).

R. H. Socolow, "Can we bury global warming," *Sci. Am.*, 293, 49–55 (July 2005).

W. Stumm and J. J. Morgan, *Aquatic Chemistry, 3rd ed.*, Wiley-Interscience, New York (1996).

J. T. Teller and D. W. Leverington "Glacial lake Agassiz: A 5000 year history of change and its relationship to the $\delta^{18}O$ record of Greenland," *Geol. Soc. Am. Bull.*, 116, 729–742 (2004).

J. R. Trabalka and D. E. Richle, eds., *The Changing Carbon Cycle: A Global Analysis*, Springer, New York (1986).

D. Walker, *Energy, Plants and Man, 2nd ed.*, Oxygraphics Ltd., Brighton (1993).

G. Walker, *Snowball Earth*, Crown, New York (2003).

J. C. G. Walker, *Evolution of the Atmosphere*, Macmillan, New York (1977).

J. B. West, *Respiratory Physiology: The Essentials, 6th ed.*, Lippincott Williams and Wilkins, Philadelphia, PA (2000).

W. B. Whitman, D. C. Coleman, and W. J. Wiebe, "Prokaryotes: The unseen majority," *Proc. Natl. Acad. Sci.*, 95, 6578–6583 (1998).

W. S. Wolbach, R. S. Lewis, and E. Anders, "Cretaceous extinctions: Evidence for wildfires and search for meteoric material," *Science*, 230, 167 (1985).

R. E. Zeebe and D. A. Wolf-Gladrow, *CO_2 in Seawater: Equilibrium, Kinetics, Isotopes*, Elsevier, Amsterdam (2001).

S. A. Zimov, S. P. Davydov, G. M. Zimova, A. I. Davydova, F. S. Chapin III, M. C. Chapin, and J. F. Reynolds, "Contribution of disturbance to increasing seasonal amplitude of atmospheric CO_2," *Science*, 284, 1973 (1999).

Index

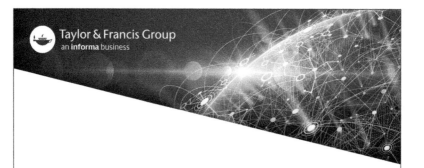